Green Heterogeneous Catalysis and Valorization Of Algerian Bentonite

Synthesis of Al-MCM-41 Mesoporous Matcrials, Lewis Acidity, and Applications in Organic Synthesis

Ali-Dahmane Tewfik

Green Heterogeneous Catalysis and Valorization Of Algerian Bentonite

Synthesis of Al-MCM-41 Mesoporous Materials, Lewis Acidity, and Applications in Organic Synthesis

ELIVA PRESS

ISBN: 978-99993-4-496-8

Cover Design: Eliva Press
Cover Image: www.ingimage.com
Email: info@elivapress.com
Website: www.elivapress.com

This Eliva Press imprint is published by the registered company Eliva Press Global Ltd. part of Eliva Press S.R.L. Publishing Group

The registered company address is: Pope Hennessy Street Level 2, Hennessy Tower Port Louis, Mauritius
Eliva Press S.R.L. Publishing Group legal address is: Bd. Cuza-Voda 1/4 of. 21, Chisinau, Moldova, Europe

PREFACE

This book was born from a deep conviction: Algerian science has important things to say to the world, and the natural resources of our country contain scientific and technological potential that remains largely untapped. It is this conviction that has guided, for more than a decade, the research work of which this book constitutes a synthesis.

When, in 2012, we undertook the exploration of the Maghnia bentonite, this natural clay occurring abundantly in the Tlemcen region and exploited for centuries for its binding and bleaching properties, as a source of functional mesoporous materials, we were following an unconventional approach. At that time, the vast majority of studies on MCM-41 relied on imported high-purity reagents that were expensive and energy-intensive to produce. The idea of synthesizing advanced nanotechnological materials from a local natural clay appeared bold, if not improbable, to some of our peers.

The results obtained largely confirmed the validity of this scientific intuition. They led to five publications in indexed international journals, the supervision of young researchers, and today to this book, which aspires to become a reference work for Algerian and international researchers, academics, and engineers working in the field of mesoporous material chemistry and green heterogeneous catalysis.

Algeria benefits from exceptional natural mineral resources. The bentonite deposit of Hammam Boughrara, in the wilaya of Tlemcen, is among the largest in the Mediterranean basin, with reserves estimated at several tens of millions of tons. Yet, this strategic resource is still mainly exported in its raw form, without significant local transformation, representing a considerable value-added opportunity that remains largely untapped.

Transforming this clay into high-technology mesoporous materials, nanomaterials with remarkable catalytic and adsorption properties, is precisely the scientific and technological challenge that this research program has addressed. This approach concretely illustrates what science can contribute to national industrial development: taking an abundant local raw material, understanding it deeply, and transforming it into high-value-added products capable of meeting growing national industrial needs in heterogeneous catalysts and efficient adsorbents for water treatment.

In writing this book, it seemed important not to limit ourselves to the description of experimental results, but rather to place these results within a broader reflection on the role of Algerian scientific research in the world and on the indispensable links between fundamental research, researcher training, and industrial development. It is in this spirit that techno-economic aspects and prospects for industrial valorization have been integrated into the structure of the book.

The environmental challenges of the twenty-first century, climate change, resource depletion, water and soil pollution, and biodiversity degradation, impose a particular responsibility on the scientific community: to design cleaner, more efficient, and less resource-intensive chemical processes.

The work presented in this book is firmly rooted in this philosophy. The synthesis of formamidine derivatives at room temperature, without solvent, using a recyclable solid catalyst derived from local bentonite, illustrates in an exemplary manner what green chemistry can achieve in practice: superior yields (93–99%) with 85% lower energy consumption, almost zero waste generation, and an environmental E-factor below 0.3. These figures speak for themselves.

Similarly, the use of MCM-41/bentonite as an adsorbent for wastewater treatment addresses an urgent national environmental need. Algerian textile and industrial effluents, often discharged without adequate treatment, constitute a serious threat to water resources. Developing efficient, low-cost, and locally produced adsorbents represents a direct and tangible contribution to environmental protection in Algeria.

This book was designed to meet the needs of several categories of readers whose interests converge around mesoporous materials, catalysis, and green chemistry. Researchers and academic staff in materials chemistry, organic chemistry, and environmental chemistry will find here a rigorous and well-referenced synthesis of original scientific results, accompanied by an in-depth critical discussion. The one hundred selected bibliographic references constitute a solid documentary basis for further exploration of each topic.

Each chapter is built around the author's original publications, explicitly cited with their complete references. This transparent organization allows the reader to return to the primary sources for deeper investigation of any point of interest. Experimental data tables are systematically accompanied by references to the publication from which they originate, thereby ensuring traceability and reproducibility of the results.

The mathematical equations used (Bragg's law, BET, BJH, Langmuir, Freundlich, adsorption kinetics, thermodynamics) are presented within the text together with their defined parameters and units, so that they may be directly applied by readers in their own work. The glossary (Appendix C) gathers the definitions of all technical terms used throughout the book.

TABLE OF CONTENTS

GENERAL INTRODUCTION

Scientific Context and Motivation

Sustainable development, green chemistry, and the valorization of local resources now constitute three priority axes of international scientific policy. In this context, the design of new functional materials from abundant and inexpensive natural sources, intended for catalytic and environmental applications, represents a major challenge in modern materials chemistry.

Ordered mesoporous materials, and in particular the MCM-41 family (Mobil Crystalline Material No. 41) discovered in 1992 [1, 2], occupy a central place in this field of research. Their unique two-dimensional hexagonal structure, combining an exceptionally high specific surface area (up to 1500 m^2/g), a narrow and uniform pore size distribution in the 2–10 nm range, and tunable surface chemistry, makes them prime candidates for numerous applications: heterogeneous catalysis [3, 4], pollutant adsorption [5, 6], controlled drug delivery [7], and molecular separation [8].

However, the classical synthesis of MCM-41 relies on the use of high-purity silica and alumina precursors (tetraethylorthosilicate, colloidal silica, sodium aluminate), whose cost and environmental constraints represent a real barrier to their large-scale industrial and applicative deployment. The search for alternative natural, economical, and ecological sources is therefore emerging as a strategic priority, particularly in the Algerian context, which is rich in natural mineral resources.

Algeria possesses one of the most significant bentonite deposits in the Mediterranean basin, located at Hammam Boughrara (Maghnia, Tlemcen province). This natural clay, composed mainly of montmorillonite and rich in silica ($SiO_2 \approx 60.5\%$) and alumina ($Al_2O_3 \approx 18.6\%$), represents an ideal aluminosilicate source for the synthesis of MCM-41, provided that an efficient method for extracting and activating the silicate and aluminate species it contains can be developed.

This is precisely the central focus of this research program: to develop an original and eco-responsible synthesis route for the preparation of Al-MCM-41 from Algerian bentonite via alkaline activation (NaOH fusion), to rigorously characterize the resulting materials, to compare their properties with those of standard materials synthesized from pure reagents, and to valorize these materials in two high societal-impact applications: green heterogeneous catalysis for the synthesis of pharmacologically relevant molecules, and wastewater treatment by adsorption.

Problem Statement and Scientific Questions

The central scientific questions that structure this work are as follows:

(1) Is it possible to synthesize Al-MCM-41 materials of satisfactory structural and textural quality using Algerian bentonite as the exclusive source of silica and aluminum, without the addition of pure precursors? What synthesis parameters (pH, temperature, bentonite/NaOH ratio) control the quality of the resulting material?

(2) What is the nature and concentration of Lewis and Brønsted acid sites generated in Al-MCM bentonite? How do they compare to those of Al-MCM-standard? What is the correlation between the aluminum content incorporated into the hexagonal phase and catalytic activity?

(3) Can Al-MCM-41 derived from bentonite constitute an efficient, recyclable, and economically attractive heterogeneous catalyst for the synthesis of formamidine derivatives under conditions consistent with the principles of green chemistry (solvent-free, ambient temperature)?

(4) Do the textural properties of MCM-bentonite (specific surface area, mesoporous volume) allow its use as a high-performance adsorbent for wastewater treatment? What kinetic and thermodynamic models govern the adsorption?

Scientific Sources of the Work

This work is based exclusively on the author's original publications in peer-reviewed international scientific journals, on pedagogical supervision work, and on complementary unpublished data. The primary sources are:

- Ali-Dahmane T. et al. (2014). *Comptes Rendus Chimie*, Elsevier [9]. Synthesis of MCM-41 from Algerian bentonite, comparison with Volclay bentonite.
- Ali-Dahmane T. et al. (2017). *Journal of Fundamental and Applied Sciences* [10]. Effect of synthesis pH on the properties of MCM-41 derived from bentonite.
- Ali-Dahmane T. et al. (2017). *Annales de Chimie Sciences des Matériaux* [11]. Control of structural properties of Si-MCM-41.
- Ali-Dahmane T. et al. (2019). *Bulletin of Chemical Reaction Engineering and Catalysis* [12]. Comparison of Lewis acidity between standard Al-MCM-41 and Al-MCM-bentonite.
- Datoussaid Y., Ali-Dahmane T. et al. (2021). *Letters in Organic Chemistry*, Bentham Science [13]. Eco-compatible synthesis of formamidines catalyzed by Al-MCM-41.
- Mesli W.M. and Merad H.Z. (2025). Master's Thesis, University of Tlemcen, supervised by the author [14]. Wastewater treatment by adsorption on MCM-41/Maghnia bentonite.

Organization of the Work

The work is structured into six main chapters, each corresponding to a specific facet of the research program:

- Chapter I presents an in-depth literature review on mesoporous materials of the M41S family and their applications.

- Chapter II describes the synthesis and characterization methods employed.
- Chapter III addresses the synthesis and control of MCM-41 properties.
- Chapter IV compares Lewis acidity between the two types of materials.
- Chapter V presents the green catalysis applications.
- Chapter VI sets out the results on wastewater treatment by adsorption.

The work concludes with a general conclusion and research perspectives.

References

1. Kresge C.T., Leonowicz M.E., Roth W.J., Vartuli J.C., Beck J.S. (1992). Ordered mesoporous molecular sieves synthesized by a liquid-crystal template mechanism. Nature, 359, 710–712.

2. Beck J.S., Vartuli J.C., Roth W.J., Leonowicz M.E., Kresge C.T., Schmitt K.D., Chu C.T.W., Olson D.H., Sheppard E.W., McCullen S.B., Higgins J.B., Schlenker J.L. (1992). A new family of mesoporous molecular sieves prepared with liquid crystal templates. *Journal of the American Chemical Society*, 114(27), 10834–10843.

3. Corma A. (1997). From microporous to mesoporous molecular sieve materials and their use in catalysis. *Chemical Reviews*, 97(6), 2373–2420.

4. Taguchi A., Schüth F. (2005). Ordered mesoporous materials in catalysis. *Microporous and Mesoporous Materials*, 77(1–2), 1–45.

5. Bui T.X., Choi H. (2009). Adsorptive removal of selected pharmaceuticals by mesoporous silica SBA-15. *Journal of Hazardous Materials*, 168(2–3), 602–608.

6. Walcarius A., Mercier L. (2010). Mesoporous organosilica adsorbents: nanoengineered materials for removal of organic and inorganic pollutants. *Journal of Materials Chemistry*, 20(22), 4478–4511.

7. Slowing I.I., Vivero-Escoto J.L., Wu C.W., Lin V.S.Y. (2008). Mesoporous silica nanoparticles as controlled release drug delivery and gene transfection carriers. *Advanced Drug Delivery Reviews*, 60(11), 1278–1288.

8. Zhao X.S., Lu G.Q., Millar G.J. (1996). Advances in mesoporous molecular sieve MCM-41. *Industrial and Engineering Chemistry Research*, 35(7), 2075–2090.

9. Ali-Dahmane T., Adjdir M., Hamacha R., Villieras F., Bengueddach A., Weidler P.G. (2014). The synthesis of MCM-41 nanomaterial from Algerian bentonite: the effect of the mineral phase contents of clay on the structure properties of MCM-41. *Comptes Rendus Chimie* (Elsevier), 17(1–2), 1–6.

10. Ali-Dahmane T., Brahmi L., Hamacha R., Villieras F., Bengueddach A. (2017). Synthesis of MCM-41 nanomaterial from Algerian bentonite: influence of synthesis pH. *Journal of Fundamental and Applied Sciences*, 9(2), 636–649.

11. Ali-Dahmane T., Brahmi L., Hamacha R., Bengueddach A. (2017). How to control the structural properties of purely siliceous MCM-41. *Annales de Chimie Sciences des Matériaux* (Lavoisier), 40(3–4), 149–163.

12. Ali-Dahmane T., Brahmi L., Hamacha R., Hacini S., Bengueddach A. (2019). Comparison of Lewis acidity between Al-MCM-41 pure chemicals and Al-MCM-41 synthesized from bentonite. *Bulletin of Chemical Reaction Engineering and Catalysis* (Scopus, Q2), 14(2), 358–368.

13. Datoussaid Y., Ali-Dahmane T., Mehiaoui N., Brahmi L., Kibou Z., Hamacha R., Choukchou-Braham N. (2021). Ecofriendly synthesis of formamidine derivatives catalyzed by Al-MCM-41 mesoporous materials. *Letters in Organic Chemistry* (Bentham Science, Scopus), 18, 1–6.

14. Mesli W.M., Merad H.Z. (2025). *Wastewater treatment by adsorption on MCM-41 synthesized from Maghnia bentonite.* Master's Thesis, Specialty: Macromolecular Chemistry, Department of Chemistry, Abou Bekr Belkaïd University of Tlemcen. Supervisor: Dr. Ali-Dahmane Tewfik.

CHAPTER I – State of the Art on Ordered Mesoporous Materials and Their Applications

I.1. Classification of Porous Materials According to IUPAC

The International Union of Pure and Applied Chemistry (IUPAC) [1] classifies porous materials into three categories based on their pore diameter: microporous materials (Dp < 2 nm), mesoporous materials (2 nm < Dp < 50 nm), and macroporous materials (Dp > 50 nm). This classification directly determines the fields of application for each category (Figure I.1).

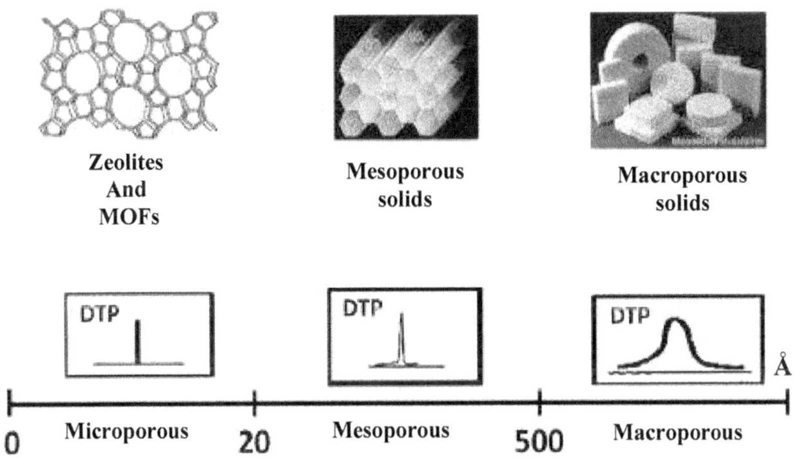

Figure I.1. Classification of porous materials according to pore size distribution [2].

Microporous materials, essentially zeolites and carbon molecular sieves, are characterized by remarkable shape selectivity, resulting from the match between pore size and substrate molecule size. They are widely used in petrochemistry, acid catalysis, and gas separation. However, their limited pore size (0.3-1.5 nm) constitutes a major obstacle for processing bulky molecules (biomolecules, polymers, drugs), which motivated the development of mesoporous materials.

Ordered mesoporous materials, of which MCM-41 is the emblematic representative, fill this gap by offering pores of controllable size between 2 and 50 nm, accessible to intermediate-sized molecules (enzymes, dyes, low molecular weight polymers, pharmaceutical active ingredients). Their large specific surface area (400-1500 m^2/g) and high porous volume (0.5-1.5 cm^3/g) are major assets for adsorption and catalysis.

I.2. M41S Family: MCM-41, MCM-48, and MCM-50

I.2.1. Discovery and Principal Members

In 1992, researchers at Mobil Oil Corporation, Beck, Kresge, Vartuli, and their collaborators, published two landmark articles [3, 4] describing the synthesis of a new family of ordered mesoporous materials, named M41S, obtained through an innovative templating mechanism using surfactant micelles as structure-directing agents. This discovery opened a new era in the chemistry of porous materials.

The M41S family comprises three principal members that differ in the geometry of their porous network: MCM-41 with a 2D hexagonal structure (space group p6mm, parallel cylindrical channels), MCM-48 with a bicontinuous cubic structure (space group Ia3d, three-dimensionally interconnected channels), and MCM-50 with a lamellar structure. MCM-41 is by far the most studied and most widely used, owing to its reproducible synthesis and exceptional textural properties (Figure I.2).

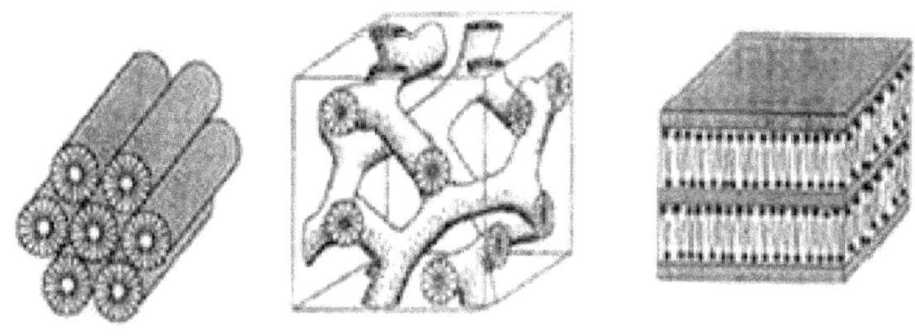

Figure I.2. The structures of MCM-41, MCM-48, and MCM-50 [5].

I.2.2. Structural and Textural Properties of MCM-41

MCM-41 exhibits a network of one-dimensional cylindrical channels, parallel to one another, arranged according to a two-dimensional hexagonal lattice. The key structural parameters are:

- The hexagonal unit cell parameter a_0 (calculated by $a_0 = 2d_{100}/\sqrt{3}$ from the (100) reflection in XRD).
- The pore diameter Dp (determined by BJH from nitrogen adsorption isotherms).
- The wall thickness $Wt = a_0 - Dp$.

For Si-MCM-41 synthesized from pure reagents with CTABr (C_{16}) as surfactant, typical values are:

- $a_0 \approx$ 40-46 Å,
- Dp $\approx$ 30-40 Å,
- Wt $\approx$ 7-12 Å,
- SBET $\approx$ 900-1500 m^2/g,

- Vmeso ≈ 0,7-1,2 cm^3/g.

These properties are highly dependent on synthesis conditions: nature and concentration of the surfactant, temperature and duration of hydrothermal treatment, pH of the medium, nature of the silica source and basic co-structuring agent [6, 7].

Table I.1. Comparison of textural and acid properties of the three types of MCM-41 materials studied in this work [8–10].

Property	Si-MCM-41 standard	Al-MCM-41 standard	Al-MCM-bentonite
SBET (m^2/g)	1154	983	494–674
Vmeso (cm^3/g)	0,84	0,76	0,42–0,72
Dp BJH (Å)	34,0	35,6	33,4–38,0
a_0 calcined (Å)	41,60	43,94	42,28–44,51
Wt (Å)	7,6	8,4	6,0–9,4
Δa_0 (%)	9,8	12,0	0,9–2,5
Si/Al (EDX)	∞ (pur Si)	13,47	43,64 (phase hexagonale)
Total acidity (mmol/g)	—	1414,77	172,77

I.2.3. Formation Mechanisms

Two principal mechanisms have been proposed to explain the formation of MCM-41-type materials. The LCT (Liquid Crystal Templating) mechanism, proposed by Beck et al. [3, 4], assumes that surfactant micelles spontaneously organize into a hexagonal liquid-crystal mesophase prior to silica condensation. The silica condenses around the hexagonal columns of micelles to form the material walls, and the surfactant is removed by calcination.

The cooperative self-assembly mechanism proposes that the hexagonal organization results from a progressive cooperative interaction between anionic silicate species and the cationic polar heads of the surfactant, leading to simultaneous growth of surfactant organization and silica condensation [11, 12]. This mechanism better accounts for experimental observations on formation kinetics and pH effects.

Several mechanistic pathways have thus been put forward, but the most widely accepted is based on a cooperative formation mechanism between mineral and organic species. It is presented in Figure I.7 for a synthesis carried out at 50°C in a dilute CTAB solution [13].

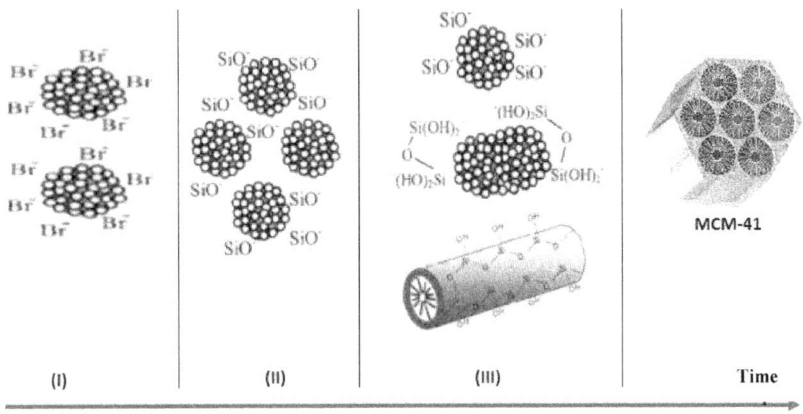

Figure I.3. Formation mechanism of MCM-41 proposed by Galarneau et al. [13].

I.3. Al-MCM-41: Aluminum Incorporation and Acidity

I.3.1. Isomorphic Si → Al Substitution

The incorporation of aluminum atoms into the siliceous framework of MCM-41 is achieved through isomorphic substitution: Si^{4+} atoms in tetrahedral coordination are replaced by Al^{3+} atoms. This substitution creates a positive charge deficit in the framework (−1 per incorporated Al), which must be compensated by a cation (H^+, Na^+, NH_4^+). When compensation is provided by H^+, a Brønsted acid site is created (Si-OH-Al bridge). When aluminum is found in an extra-framework position (tetrahedral or octahedral coordination), Lewis acid sites are generated.

The concentration of Lewis (CL) and Brønsted (CB) acid sites is directly correlated to the amount of incorporated aluminum, expressed by the Si/Al ratio. The lower this ratio (high Al content), the higher the concentration of acid sites. Ali-Dahmane et al. [10] demonstrated that for Al-MCM-standard (Si/Al = 13.47), the total concentration of acid sites is 8 times higher than that of Al-MCM-bentonite (Si/Al = 43.64 in the hexagonal phase), which translates into marked differences in catalytic performance in test reactions.

I.3.2. Acidity Characterization by FTIR-Pyridine

FTIR spectroscopy coupled with pyridine adsorption is the reference method for identifying and quantifying Lewis and Brønsted acid sites in aluminosilicate materials. Pyridine (probe molecule with kinetic diameter ≈ 5.7 Å, compatible with MCM-41 mesopores of diameter 35–38 Å) forms characteristic complexes with the different types of acid sites:

- Pyridinium (Py-B): formed with Brønsted sites, bands at 1545 cm^{-1} and 1636 cm^{-1}.
- Coordinated pyridine (Py-L): complexes with Lewis sites, bands at 1456 cm^{-1} and 1622 cm^{-1}.
- Physisorbed / hydrogen-bonded pyridine (Py-H): bands at 1445 cm^{-1} and 1596 cm^{-1}.
- Mixed band (Py-B+L): 1490 cm^{-1}.

The concentrations CL and CB are calculated by integration of the bands at 1456 and 1545 cm^{-1} respectively, using the Emeis extinction coefficients [14]: $\varepsilon B = 0.73$ $cm \cdot \mu mol^{-1}$ and $\varepsilon L = 1.11$ $cm \cdot \mu mol^{-1}$. Progressive desorption of pyridine at increasing temperatures (50, 150, 250, and 350°C) allows the strength of acid sites to be assessed: strong sites retain pyridine at high temperature.

I.4. Synthesis of MCM-41 from Algerian Bentonite

I.4.1. Maghnia Bentonite: An Algerian Strategic Resource

Maghnia bentonite (Hammam Boughrara, Tlemcen province) is exploited by the National Enterprise for Non-Ferrous Mineral Products (ENOF) and represents one of the most significant bentonite reserves in North Africa, estimated at several tens of millions of tonnes. Its chemical composition, determined by X-ray fluorescence [9], reveals a predominant content of SiO_2 (60.49%) and Al_2O_3 (18.59%), with a natural Si/Al ratio of 4.76, ideal for the synthesis of aluminosilicate materials.

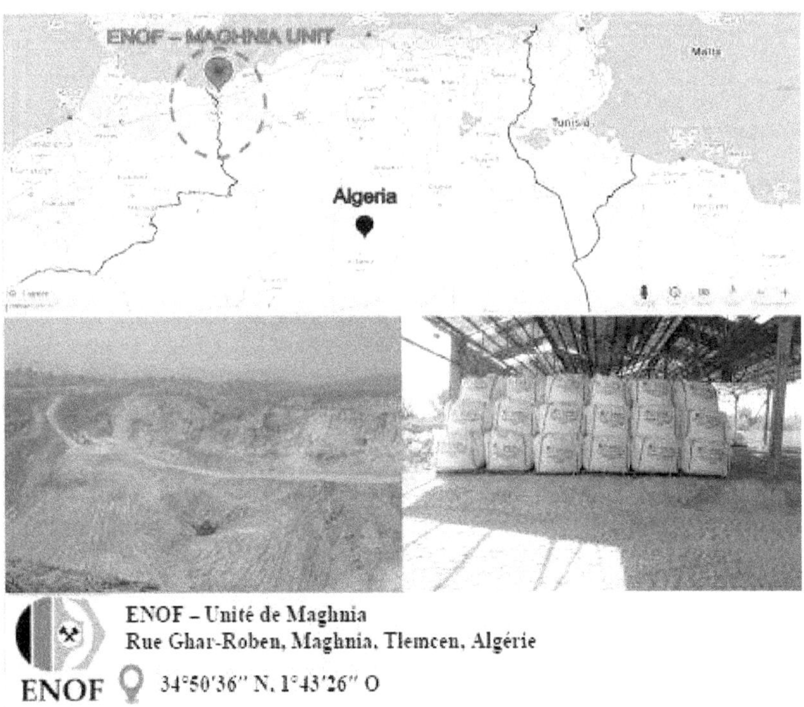

Figure I.4. Location of the Maghnia bentonite deposit.

The mineralogy of the bentonite is dominated by montmorillonite (2:1 phyllosilicate, d_{001} = 12.65 Å, Figures I.5 and I.6) as the principal clay mineral, accompanied by kaolinite (d = 7.16 Å), illite (d = 10 Å), and quartz (d = 4.25 Å and 3.34 Å) as non-clay minerals.

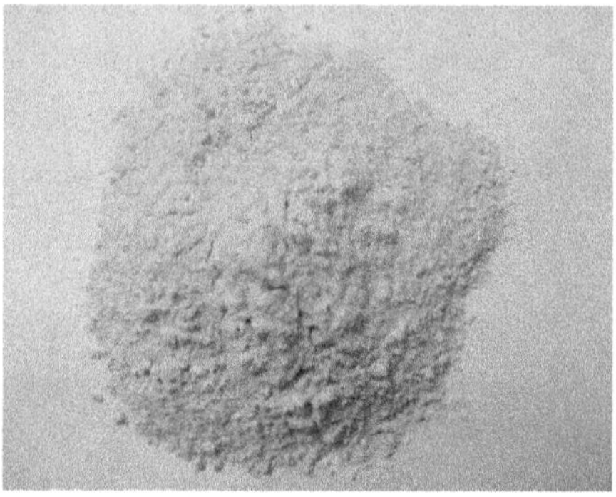

Figure II.1. Bentonite of Maghenia.

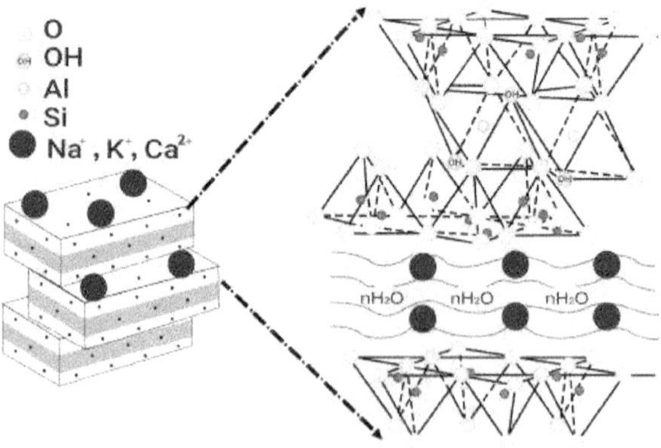

Figure I.6. Structure of the montmorillonite [15].

These phases disappear entirely upon alkaline fusion at 550°C, confirming the complete dissolution of the crystalline structure of the clay.

I.4.2. Valorization Strategy by Alkaline Fusion

The strategy developed by Ali-Dahmane and al. [8, 9] consists of selectively extracting silicate and aluminate species from bentonite through alkaline activation, by mixing powdered bentonite with NaOH in a 1:1.2 mass ratio, then calcining the mixture at 550°C for 1 hour. The fused solid is dissolved in demineralized water (1:4 ratio), and the supernatant obtained by centrifugation is used directly as an aluminosilicate precursor for the hydrothermal synthesis of MCM-41 in the presence of CTABr.

This approach offers the considerable advantage of completely avoiding the use of pure silica and alumina sources (TEOS, Ludox colloidal silica, sodium aluminate), which reduces raw material costs by approximately 50% and significantly improves the environmental footprint of the process. It is consistent with the philosophy of green chemistry and sustainable development, in line with current national and international scientific priorities.

I.4.3. Comparison with Other Natural Sources in the Literature

The valorization of natural sources for the synthesis of MCM-41 has attracted growing interest in the international scientific community. Table I.2 presents a comparative summary of the principal works published in this field.

Table I.2. Comparative summary of MCM-41 materials prepared from various natural sources.

Natural source	Region	SBET (m^2/g)	Si/Al	Extraction method	Reference
Maghnia bentonite	Algeria	494	43.64	NaOH alkaline fusion	[8, 9]
Volclay bentonite	Germany	1060	18.5	NaOH alkaline fusion	[16]
Bentonite (China)	China	1018	36	NaOH alkaline fusion	[17]
Halloysite	China	580	—	NaOH dissolution	[18]
Kaolin	China	720	—	Metakaolinization	[19]
Diatomite	China	650	—	NaOH attack	[20]

Coal fly ash	USA	842	13.4	Alkaline fusion	[21]
Rice husk ash	Brazil	910	—	Alkaline extraction	[22]
K-Feldspar	China	560	—	HF/NaOH fusion	[23]

I.5. Applications of MCM-41 Materials

I.5.1. Heterogeneous Catalysis in Organic Synthesis

Al-MCM-41 is a versatile acid catalyst whose accessible surface enables it to catalyze a wide variety of organic reactions. Brahmi and al. [24] demonstrated that standard Al-MCM-41 effectively catalyzes the allylation of benzaldehyde with allyltrimethylsilane at 35°C in CH_2Cl_2, with a yield of 99% in 4 hours, even surpassing homogeneous $TiCl_4$ in terms of environmental benefits (recoverable and recyclable heterogeneous catalyst). This model reaction has become the standard test reaction for evaluating the Lewis acidity of mesoporous materials in our research group.

Datoussaid and Ali-Dahmane and al. [25] extended the application of Al-MCM-41 to the synthesis of formamidine derivatives from primary amines and DMF-DMA, under solvent-free conditions at room temperature. This original contribution demonstrates that the low-to-moderate intensity Lewis acid sites present in Al-MCM-bentonite are sufficient to activate this reaction, thus opening the way for the use of less acidic but more economical materials for certain catalytic applications.

I.5.2. Adsorption and Water Treatment

The exceptional textural properties of MCM-41, large specific surface area (494-1154 m^2/g), uniform mesoporous channels (Dp = 33-38 Å), tunable surface chemistry, make it an adsorbent of choice for the removal of organic and inorganic pollutants from wastewater. Mesli, Merad, and Ali-Dahmane [26] demonstrated the effectiveness of MCM-41 synthesized from Maghnia bentonite for the adsorption of cationic dyes, with fast kinetics (equilibrium < 60 min) and satisfactory adsorption capacity.

The strategic advantage of MCM-bentonite for this application lies in its low production cost (local bentonite + NaOH + CTABr) compared to conventional adsorbents such as activated carbon or synthetic zeolites, making it economically attractive for the treatment of Algerian industrial effluents.

I.6. Green Chemistry and Sustainable Development

I.6.1. The 12 Principles of Green Chemistry

Green chemistry, whose 12 founding principles were formulated by Anastas and Warner in 1998 [27], provides a conceptual framework for designing sustainable chemical processes. The work presented in this book respects and exemplifies several of these principles in an exemplary manner:

- **Principle 1 — Prevention:**

  The synthesis of formamidines under solvent-free conditions eliminates the generation of organic solvent waste. The E-factor (waste mass/product mass ratio) calculated for this reaction is below 0.15, remarkably low for fine chemistry ($E > 5$ typically).

- **Principle 6 — Energy Efficiency:**

  Reaction conditions at room temperature (vs. reflux at 130–140°C for the classical thermal method) reduce energy consumption by approximately 85%, representing a significant environmental and economic benefit at industrial scale.

- **Principle 9 — Catalysis:**

  The use of a recyclable solid heterogeneous catalyst (Al-MCM-41) in place of stoichiometric reagents or toxic homogeneous Lewis acids (BF_3, $TiCl_4$, $AlCl_3$) is the core of the green approach developed.
- **Principle 10 — Degradability:**

Al-MCM-41 materials are inorganic oxides (silica and alumina) that are perfectly stable, non-toxic, and have no negative environmental impact at end of life.

I.6.2. Atom Economy of Formamidine Synthesis

Atom economy (AE), defined by Trost [28] as the ratio between the molar mass of the desired product and the sum of the molar masses of all reagents, is a key efficiency indicator in green chemistry.

For the synthesis of formamidines:

$$AE = MM(formamidine) / [MM(amine) + MM(DMF\text{-}DMA)] \times 100\%$$

Calculated for formamidine **2h** (MM = 236 g/mol):

$$AE = 236 / (127 + 119) \times 100 = \mathbf{96\%}$$

This exceptionally high value confirms that the reaction is near-quantitative in terms of incorporation of reagent atoms into the final product, with only 2 molecules of methanol (CH_3OH, MM = 32 g/mol) as by-products.

I.7. Conclusion of Chapter I

This literature review has established the conceptual framework and state of the art necessary for understanding the original work presented in the following chapters. The essential points to retain are:

(i) the central place of MCM-41 among mesoporous materials and its remarkable structural properties;

(ii) (ii) the decisive role of aluminum on surface acidity and catalytic properties;

(iii) the relevance and originality of the valorization of Algerian Maghnia bentonite;

(iv) the diversity of valorizing applications (green catalysis, water treatment);

(v) the grounding in the principles of green chemistry and sustainable development.

I.8. References of Chapter I

1. Sing K.S.W., Everett D.H., Haul R.A.W., Moscou L., Pierotti R.A., Rouquerol J., Siemieniewska T. (1985). Reporting physisorption data for gas/solid systems with special reference to the determination of surface area and porosity (Recommendations 1984). *Pure and Applied Chemistry*, 57(4), 603–619.

2. Moukahhal K. (2020). *Synthèse et mise en forme d'objets poreux hiérarchisés composés de nanozéolithes*. Ph.D. Dissertation, Université de Haute Alsace-Mulhouse, France.

3. Kresge C.T., Leonowicz M.E., Roth W.J., Vartuli J.C., Beck J.S. (1992). Ordered mesoporous molecular sieves synthesized by a liquid-crystal template mechanism. *Nature*, 359, 710–712.

4. Beck J.S., Vartuli J.C., Roth W.J., Leonowicz M.E., Kresge C.T., Schmitt K.D., Chu C.T.W., Olson D.H., Sheppard E.W., McCullen S.B., Higgins J.B., Schlenker J.L. (1992). A new family of mesoporous molecular sieves prepared with liquid crystal templates. *Journal of the American Chemical Society*, 114(27), 10834–10843.

5. Bhattacharyya S., Lelong G., Saboungi M.L. (2006). Recent progress in the synthesis and selected applications of MCM-41: a short review. *Journal of Experimental Nanoscience*, 1(3), 375–395.

6. Ali-Dahmane T., Brahmi L., Hamacha R., Bengueddach A. (2017). How to control the structural properties of purely siliceous MCM-41. *Annales de Chimie Sciences des Matériaux*, 40(3–4), 149–163.

7. Cestors Y., Haller G.L. (2001). Several factors affecting Al-MCM-41 synthesis. *Microporous and Mesoporous Materials*, 43(3), 171–179.

8. Ali-Dahmane T., Adjdir M., Hamacha R., Villieras F., Bengueddach A., Weidler P.G. (2014). The synthesis of MCM-41 nanomaterial from Algerian bentonite: the effect of the mineral phase contents of clay on the structure properties of MCM-41. *Comptes Rendus Chimie*, 17(1–2), 1–6.

9. Ali-Dahmane T., Brahmi L., Hamacha R., Villieras F., Bengueddach A. (2017). Synthesis of MCM-41 nanomaterial from Algerian bentonite: influence of synthesis pH. *Journal of Fundamental and Applied Sciences*, 9(2), 636–649.

10. Ali-Dahmane T., Brahmi L., Hamacha R., Hacini S., Bengueddach A. (2019). Comparison of Lewis acidity between Al-MCM-41 pure chemicals and Al-MCM-41 synthesized from bentonite. *Bulletin of Chemical Reaction Engineering and Catalysis*, 14(2), 358–368.

11. Firouzi A., Kumar D., Bull L.M., Besier T., Sieger P., Huo Q., Walker S.A., Zasadzinski J.A., Glinka C., Nicol J., Margolese D., Stucky G.D., Chmelka B.F. (1995). Cooperative organization of inorganic-surfactant and biomimetic assemblies. *Science*, 267(5201), 1138–1143.

12. Vartuli J.C., Schmitt K.D., Kresge C.T., Roth W.J., Leonowicz M.E., McCullen S.B., Hellring S.D., Beck J.S., Schlenker J.L., Olson D.H., Sheppard E.W. (1994). Effect of surfactant/silica molar ratios on the formation of mesoporous molecular sieves. *Chemistry of Materials*, 6(12), 2317–2326.

13. Galarneau A., Di Renzo F., Fajula F., Mollo L., Fubini B., Ottaviani M.F. (1998). Kinetics of formation of micelle-templated silica mesophases monitored by electron paramagnetic resonance. *Journal of Colloid and Interface Science*, 201, 105–117.

14. Emeis C.A. (1993). Determination of integrated molar extinction coefficients for infrared absorption bands of pyridine adsorbed on solid acid catalysts. *Journal of Catalysis*, 141(2), 347–354.

15. Ubowska A. (2010). Montmorillonite as a polyurethane foams flame retardant. *Archivum Combustionis*, 30(4), 459–462.

16. Adjdir M., Ali-Dahmane T., Weidler P.G., Friedrich F., Scherer T. (2009). The synthesis of Al-MCM-41 from volclay: a low-cost Al and Si source. *Applied Clay Science*, 46(2), 185–189.

17. Yang H., Deng Y., Du C., Jin S. (2010). Novel synthesis of ordered mesoporous materials Al-MCM-41 from bentonite. *Applied Clay Science*, 47(3–4), 351–355.

18. Xie Y.L., Zhang Y., Ouyang J., Yang H.M. (2014). Mesoporous material Al-MCM-41 from natural halloysite. *Physics and Chemistry of Minerals*, 41(7), 497–503.

19. Du C.F., Yang H.M. (2012). Investigation of the physicochemical aspects from natural kaolin to Al-MCM-41 mesoporous materials. *Journal of Colloid and Interface Science*, 369(1), 216–222.

20. Yu Z.H., Wang Y., Liu X.Y., Sun J.B., Sha G.Y., Yang J.H., Meng C.G. (2014). A novel pathway for the synthesis of ordered mesoporous silica from diatomite. *Materials Letters*, 119, 150–153.

21. Kumar P., Mal N., Oumi Y., Yamana K., Sano T. (2001). Mesoporous materials prepared using coal fly ash as the silicon and aluminium source. *Journal of Materials Chemistry*, 11(12), 3285–3290.

22. Wang W.X., Martin J.C., Fan X.T., Han A.J., Luo Z.P., Sun L.Y. (2012). Silica nanoparticles and frameworks from rice husk biomass. *ACS Applied Materials and Interfaces*, 4(2), 977–981.

23. Miao S., Liu Z., Ma H., Han B., Du J., Sun Z., Miao Z. (2005). Synthesis and characterization of mesoporous aluminosilicate molecular sieve from K-feldspar. *Microporous and Mesoporous Materials*, 83(1–3), 277–282.

24. Brahmi L., Ali-Dahmane T., Hamacha R., Hacini S. (2016). Catalytic performance of Al-MCM-41 catalyst for the allylation of aromatic aldehydes with allyltrimethylsilane: comparison with $TiCl_4$ as Lewis acid. *Journal of Molecular Catalysis A: Chemical*, 423, 31–40.

25. Datoussaid Y., Ali-Dahmane T., Mehiaoui N., Brahmi L., Kibou Z., Hamacha R., Choukchou-Braham N. (2021). Ecofriendly synthesis of formamidine derivatives catalyzed by Al-MCM-41 mesoporous materials. *Letters in Organic Chemistry*, 18, 1–6.

26. Mesli W.M., Merad H.Z. (2025). *Wastewater treatment by adsorption on MCM-41 synthesized from Maghnia bentonite*. Master's Thesis, Specialty: Macromolecular Chemistry, Department of Chemistry, Abou Bekr Belkaïd University of Tlemcen. Supervisor: Dr. Ali-Dahmane Tewfik.

27. Anastas P.T., Warner J.C. (1998). *Green Chemistry: Theory and Practice*. Oxford University Press, New York, USA.

28. Trost B.M. (1991). The atom economy: a search for synthetic efficiency. *Science*, 254(5037), 1471–1477.

CHAPTER II – Materials, Synthesis Protocols and Characterization Techniques

II.1. Raw Materials and Reagents

II.1.1. Chemical Reagents for Synthesis

All chemical reagents used in this work were employed without prior purification. The following analytical or technical grade reagents were used:

- Cetyltrimethylammonium bromide (CTABr, Merck, purity ≥ 99%), the principal surfactant agent for MCM-41 structuring.

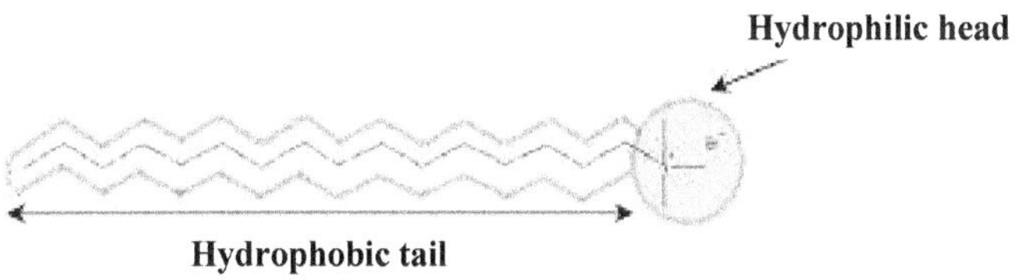

Figure II.1. Structure of cetyltrimethylammonium bromide (CTAB).

- Colloidal silica (Ludox 40%, Prolabo), silica source for the reference synthesis.
- Tetramethylammonium hydroxide pentahydrate ($TMAOH \cdot 5H_2O$, 97%, Aldrich), co-structuring agent and base.
- Sodium aluminate (54% Al_2O_3, 41% Na_2O, 5% H_2O, Aldrich), aluminum source for standard Al-MCM-41.
- Ammonium hydroxide (25%, Aldrich).
- Glacial acetic acid (99.8%, Aldrich) for pH adjustment.

For catalytic reactions:

- N,N-Dimethylformamide dimethyl acetal (DMF-DMA, 97%, Aldrich).
- Benzaldehyde (99%, Aldrich); allyltrimethylsilane (97%, Aldrich); dichloromethane (anhydrous solvent, Aldrich).

- Aromatic and heteroaromatic primary amines used for formamidine synthesis (Aldrich, purity ≥ 95–99%).

II.1.2. Natural Bentonite from Maghnia

The bentonite was used as received, without any prior treatment other than air drying and grinding to obtain a fine powder (< 100 μm).

- Chemical composition (wt%):
  SiO_2 = 60.49; Al_2O_3 = 18.59; Fe_2O_3 = 2.29; MgO = 3.80; TiO_2 = 0.10; CaO = 1.00; Na_2O = 1.20; K_2O = 0.90; loss on ignition = 11.63.

- The calculated Si/Al ratio is 4.76.

- Mineralogical phases identified by XRD: montmorillonite (major mineral, d_{001} = 12.65 Å), kaolinite (d = 7.16 Å), illite (d = 10 Å), and quartz (d = 4.25 Å).

II.2. Synthesis Protocols

II.2.1. Synthesis of Standard Si-MCM-41 (Figure II.1

The molar composition of the synthesis gel for standard Si-MCM-41 is:

$$1\ SiO_2 : 0.25\ CTABr : 0.2\ TMAOH : 40\ H_2O.$$

Experimental procedure:

- (1) Dissolve CTABr (0.867 g) in an aqueous solution of TMAOH (15 mL, 0.2 M solution) under magnetic stirring for 30 minutes at room temperature (R.T.).

- (2) Add the silica source (Ludox silica, fumed silica, or TEOS depending on the experiment) dropwise under vigorous stirring.
- (3) Stir for 2 h at R.T.
- (4) Transfer to a Teflon autoclave and treat hydrothermally (80–150°C, 24–72 h).
- (5) Filter, wash thoroughly with water, dry for 24 h at 100°C
.
- (6) Calcine for 12 h at 550°C in air (ramp: 2°C/min). The calcined material is designated Si-MCM-standard.

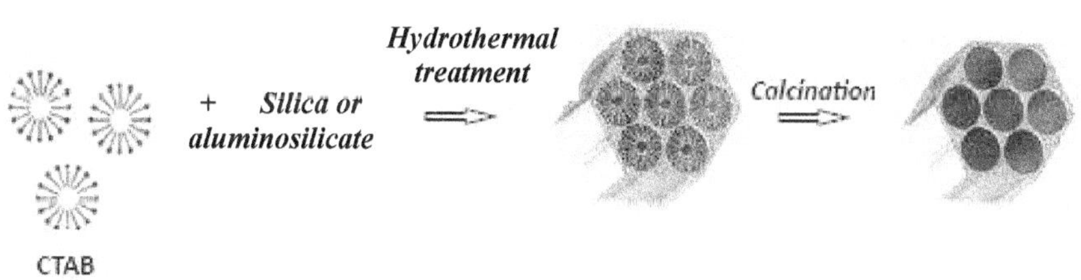

Figure II.2. Scheme of the synthesis of MCM-41.

II.2.2. Synthesis of Standard Al-MCM-41

Molar composition:

$$1\ SiO_2 : 0.25\ CTABr : 0.2\ TMAOH : 0.04\ Al_2O_3 : 40\ H_2O$$

(Theoretical Si/Al = 12.5).

Sodium aluminate is dissolved in the TMAOH solution prior to the addition of colloidal silica. Steps (4)–(6) are identical to Si-MCM-41, with hydrothermal treatment at 100°C for 48 hours. The calcined material is designated Al-MCM-standard.

II.2.3. Synthesis of MCM-41 from Bentonite (MCM-bentonite)

Step 1 — Alkaline activation of bentonite (Figure II.3):

Mix bentonite (10 g) and NaOH (12 g) in a 1:1.2 mass ratio. Calcine the mixture at 550°C for 1 hour in air in an alumina crucible. Cool to R.T.

Figure II.3. Bentonite/NaOH mixture after heating for 3 hours at 550 °C.

After cooling the bentonite/NaOH mixture to room temperature, the obtained solid is ground (Figure II.4).

Figure II.4. The mixture after grinding.

Step 2 — Supernatant extraction (Figures II.5 and II.6):

Add 40 mL of demineralized water to the fused solid, stir for 24 h at R.T. Centrifuge at 4000 rpm for 20 minutes. Recover the clear supernatant (basic solution, pH ≈ 12–13).

Figure II.5 Mixture under stirring

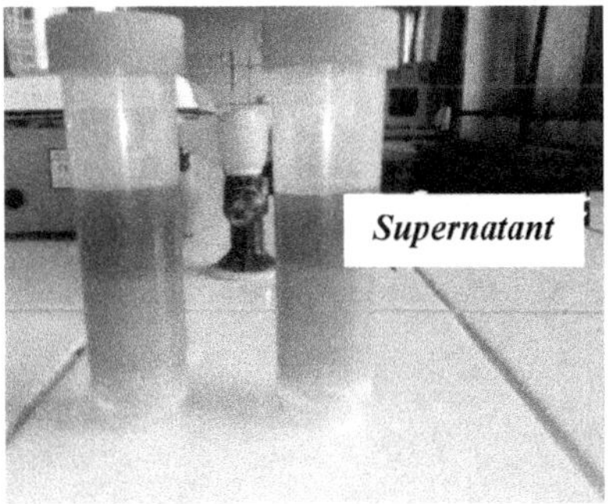

Figure II.6. The supernatant.

Step 3 — Hydrothermal synthesis:

- Mix 40 mL of the supernatant with CTABr (0.867 g), NH_4OH (0.75 g), and 15 mL of distilled water. Stir for 1 h at R.T.
- Adjust pH to the desired value (8.5–11.5) with 1 M AcOH.
- Treat hydrothermally at the chosen temperature for 48 h.

Step 4 — Post-treatment:

- Filter, wash thoroughly.
- Dry at 100°C (24 h).
- After the hydrothermal step, the (CTA^+)-MCM-41 solid (Figure II.7) is calcined in air at 550°C for 12 hours (2°C/min) to extract the structure-directing agent and empty the pores.

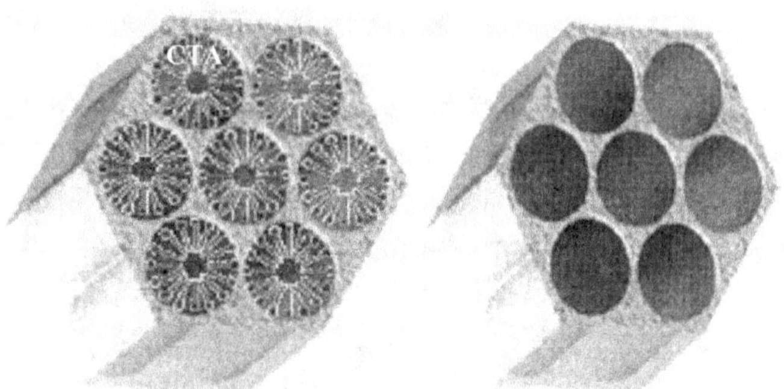

Figure II. 7. Simplified scheme of CTA^+ extraction by calcination.

II.3. Characterization Techniques

II.3.1. X-Ray Diffraction (XRD)

- Instrument: Bruker D5005 diffractometer.

- Cu-Kα anticathode (λ = 1.54056 Å) or Co-Kα (λ = 1.78897 Å).
- Conditions: angular range 2θ = 1°–10° (small angles, mesoporous structure) and 5°–70° (wide angles, crystalline phases).
- Step size 0.02°, counting time 2 s/step.
- The d_{100} spacing is calculated using Bragg's law ($n\lambda = 2d \sin \theta$).
- Unit cell parameter $a_0 = 2d_{100}/\sqrt{3}$.

II.3.2. Nitrogen Adsorption-Desorption at 77 K

- Instruments: Micromeritics GEMINI 2375 or Nova 1000e (Quantachrome).
- Preparation: outgassing under vacuum at 200°C for 24 hours.
- BET domain: $0.04 \leq P/P_0 \leq 0.20$.
- Pore size distribution: BJH method applied to the desorption branch.
- Calculated textural parameters:

SBET (m^2/g), Vmeso (cm^3/g), Dp (Å), and $Wt = a_0 - Dp$ (Å).

II.3.3. FTIR Spectroscopy

- Spectrometer: Nicolet Avatar 360 FT-IR.
- KBr pellets (1–2 mg sample / 200 mg KBr).
- Resolution: 4 cm^{-1}, 64 scans.

- For acidity study: in situ pyridine adsorption after outgassing at 450°C under secondary vacuum (10^{-3} Torr, 12 h). Spectra recorded after progressive desorption at 50, 150, 250, and 350°C under 3 Torr vacuum. Calculation of CL and CB concentrations by integration of bands at 1456 and 1545 cm^{-1} (Emeis coefficients [1]).

II.3.4. Electron Microscopy

- SEM: Philips ESEM XL 30 FEG, 15–20 kV, gold-coated samples.

- TEM: Philips CM20 with EDX analyzer (Philips EDXS), 200 kV. TEM preparation: dispersion in ethanol, deposition on perforated formvar carbon grid. Point EDX analysis in hexagonal and amorphous regions allows determination of the local Si/Al ratio.

II.3.5. Thermal Analysis (DTA/TGA)

- Instrument: Setaram.

- Conditions: temperature ramp of 10°C/min under dry air flow (50 mL/min), R.T. → 900°C.

- This analysis identifies thermal events associated with the decomposition of the organic surfactant (200–600°C), the departure of structural water, and the condensation of wall silanols.

II.3.6. UV-Visible Spectroscopy for Monitoring Catalytic and Adsorption Reactions

- Spectrophotometer: Shimadzu UV-1800 or equivalent.
- For catalytic reactions: monitoring by TLC and yield determination by NMR after purification.
- For adsorption: measurement of residual pollutant concentration at their maximum absorption wavelength (λmax).

II.4. Catalytic Test Protocols

II.4.1. Synthesis of Formamidine Derivatives

General procedure (after Datoussaid and Ali-Dahmane et al. [2]):

- In a dried flask, mix the primary amine (1 mmol), DMF-DMA (1.2 mmol), and Al-MCM-41 (10 wt%) at room temperature, without solvent.
- Stir until complete consumption of the starting material (TLC, eluent: PE/EtOAc 3/1). Add absolute EtOH (10 mL).
- Filter the catalyst.
- Concentrate the filtrate to 2 mL to induce product precipitation.
- Filter, wash with absolute EtOH.
- For recycling: wash the catalyst (3 × EtOH), dry for 3 h at 100°C, reuse directly.

II.4.2. Allylation of Benzaldehyde

Procedure (after Brahmi and Ali-Dahmane et al. [3]):

- In a two-necked flask under argon, mix benzaldehyde (1 mmol, 0.106 g), CH_2Cl_2 (2 mL), catalyst (25 mol%).
- Stir for 5 min at 35°C.
- Slowly add allyltrimethylsilane (1.5–3 equiv.) in CH_2Cl_2 (0.5 mL) at constant 35°C.
- Monitor by TLC (PE/EtOAc 20/1).

- Filter the catalyst.
- Purify by column chromatography on silica (PE/EtOAc 28/1).
- Characterize by ^{1}H NMR and ^{13}C NMR.

II.5. Adsorption Test Protocols

Solution preparation:

- Dissolve the pollutant (methylene blue) in distilled water at the desired initial concentration C_0 (10–200 mg/L).
- Adjust pH to the desired value with dilute HCl or NaOH.

Adsorption test:

- Weigh 10–50 mg of MCM-bentonite into an Erlenmeyer flask.
- Add 50–100 mL of pollutant solution.
- Stir (300 rpm) in a thermostated bath at the desired temperature (20–60°C).
- Withdraw aliquots at defined time intervals.
- Filter through membrane (0.45 µm).
- Measure residual concentration by UV-Vis. Calculate $q_e = (C_0 - C_e) \times V / m$.

Modeling:

- Isotherms are fitted to the Langmuir [4] and Freundlich [5] models.

- Kinetics are modeled using the pseudo-first-order (Lagergren [6]) and pseudo-second-order (Ho and McKay [7]) equations.

- Thermodynamic parameters ($\Delta G°$, $\Delta H°$, $\Delta S°$) [5, 8] are determined by plotting $\ln(Kd) = f(1/T)$.

II.6. References of Chapter II

1. Emeis C.A. (1993). Determination of integrated molar extinction coefficients for infrared absorption bands of pyridine adsorbed on solid acid catalysts. *Journal of Catalysis*, 141(2), 347–354.

2. Datoussaid Y., Ali-Dahmane T., Mehiaoui N., Brahmi L., Kibou Z., Hamacha R., Choukchou-Braham N. (2021). Ecofriendly synthesis of formamidine derivatives catalyzed by Al-MCM-41 mesoporous materials. *Letters in Organic Chemistry*, 18, 1–6.

3. Brahmi L., Ali-Dahmane T., Hamacha R., Hacini S. (2016). Catalytic performance of Al-MCM-41 catalyst for the allylation of aromatic aldehydes with allyltrimethylsilane: comparison with $TiCl_4$ as Lewis acid. *Journal of Molecular Catalysis A: Chemical*, 423, 31–40.

4. Langmuir I. (1918). The adsorption of gases on plane surfaces of glass, mica and platinum. *Journal of the American Chemical Society*, 40, 1361–1403.

5. Abd El-Rahman K.M., El-Kamash A.M., El-Sourougy M.R., Abdel-Moniem N.M. (2006). *Journal of Radioanalytical and Nuclear Chemistry*, 268, 221–223.

6. Lagergren S. (1898). About the theory of so-called adsorption of solution substances. *24*(4), 1–39.

7. Ho Y.S., McKay G. (1999). Pseudo-second order model for sorption processes. *Process Biochemistry*, 34, 451–465.

8. Van't Hoff J.H. (1884). *Études de Dynamique Chimique*. Amsterdam: Frederik Muller & Co.

CHAPTER III – Synthesis, Control and Optimization of the Properties of MCM-41

III.1. Control of the Structural Properties of Si-MCM-41

III.1.1. Original contribution: how to control the properties of MCM-41

Ali-Dahmane and al. published in 2017 in *Annales de Chimie Sciences des Matériaux* [1] a systematic study on the control of the structural and textural properties of purely siliceous Si-MCM-41 synthesized from high-purity reagents. This work represents the reference contribution of the research group, establishing structure, synthesis relationships for standard MCM-41 prior to addressing synthesis from natural sources.

The study investigated three main variables:

- The nature of the silica source (Ludox colloidal silica, silica fume, TEOS)
- The nature of the basic co-structure directing agent (NaOH, TMAOH, TEAOH)
- Temperature (80–150°C)
- Hydrothermal treatment time (12–72 h)

III.1.2. Effect of silica source on structural ordering

Low-angle XRD patterns of Si-MCM-41 materials synthesized from the three silica sources at 100°C in TMAOH medium reveal significant differences in structural ordering.

Ludox colloidal silica leads to the most highly ordered materials, characterized by an intense and narrow (100) peak and the presence of (110), (200), and (210) reflections, indicating a high-quality long-range hexagonal organization.

Silica fume gives slightly lower results, while TEOS leads to materials of intermediate order but with slightly larger pores, due to slower hydrolysis kinetics.

These differences are attributed to the reactivity and initial polymerization state of silicate species in solution: colloidal silica, already present as silicate oligomers in alkaline solution, reacts more rapidly and in a more controlled manner with CTABr surfactant, leading to a more ordered material.

Tableau III.1. Textural properties of Si-MCM-41 synthesized with different silica sources, basic agents, and temperatures. Data from Ali-Dahmane and al. [1].

SiO_2 source	Basic agent	T (°C)	SBET (m^2/g)	Vmeso (cm^3/g)	Dp (Å)	a_0 (Å)	Hexagonal order
Ludox	TMAOH	100	1210	0,88	35,2	43,1	Very good
Ludox	NaOH	100	1154	0,84	34,0	41,6	Very good
Ludox	TEAOH	100	1175	0,85	34,8	42,8	Good
Silica fume	TMAOH	100	1090	0,79	34,2	41,8	Good
Silica fume	NaOH	100	1040	0,76	33,5	40,9	Good
TEOS	TMAOH	100	1080	0,79	36,8	44,2	Medium-good
TEOS	NaOH	100	1020	0,74	36,1	43,5	Medium
Ludox	TMAOH	80	1050	0,72	32,8	40,1	Good
Ludox	TMAOH	130	1080	0,77	34,5	42,3	Good
Ludox	TMAOH	150	960	0,68	33,9	41,6	Medium

III.1.3. Effect of the basic co-structure directing agent

Among the three basic agents tested, TMAOH systematically leads to the most highly ordered materials and the highest specific surface areas, regardless of silica source. This effect is attributed to the dual role of TMAOH: (i) a basic agent promoting silica hydrolysis, and (ii) a co-surfactant, where TMA^+ cations (tetramethylammonium) can intercalate between CTABr head groups and stabilize the hexagonal micellar organization.

NaOH gives comparable results to TMAOH for Ludox silica but poorer results for TEOS, likely because Na^+ perturbs CTABr organization less effectively than TMA^+.

TEAOH, whose TEA^+ cations are bulkier than TMA^+, may hinder micellar organization and thus yields intermediate results.

III.1.4. Effect of crystallization temperature

The optimal crystallization temperature is 100°C, leading to materials with the highest specific surface area (1154–1210 m^2/g) and best structural ordering. At 80°C, silica condensation kinetics are slower, producing slightly less ordered but still acceptable materials. Above 130°C, a slight degradation of structural order and surface area is observed, attributed to partial dissolution of silica in alkaline medium at high temperature followed by less ordered reprecipitation.

The optimal hydrothermal treatment time is 36–48 h at 100°C. Samples collected before 24 h show incomplete structural ordering (weak or absent higher-order peaks), while beyond 48 h the properties stabilize.

III.2. Synthesis of MCM-41 from Algerian Bentonite

III.2.1. Foundational results: Comptes Rendus Chimie 2014 article

The first demonstration of MCM-41 synthesis from Algerian Maghnia bentonite was reported by Ali-Dahmane et al. in 2014 in *Comptes Rendus Chimie* (Elsevier) [2]. This pioneering paper presents a systematic comparison between MCM-41 derived from Maghnia bentonite and MCM-41 derived from Volclay bentonite (international commercial reference), both synthesized under identical conditions.

XRD results for both materials confirm the formation of the hexagonal MCM-41 phase, evidenced by the characteristic (100) peak at $2\theta \approx 2.27°$ for MCM-bentonite (Maghnia) and $2\theta \approx 2.20°$ for MCM-Volclay. However, important differences are observed: MCM-bentonite exhibits higher-order reflections (110), (200), (210), and even (300), indicating a remarkably well-developed hexagonal organization, superior to that of standard Al-MCM prepared from pure reagents with identical Si/Al ratio.

This unexpected result, better structural ordering for the bentonite-derived material compared to the standard one, was rationalized by the presence of natural cations (Ca^{2+}, Mg^{2+}, Na^{+}) in the molten bentonite supernatant, which may act as cross-linking agents for silicate species around CTABr micelles, promoting a more regular organization.

III.2.2. Synthesis pH optimization

The 2017 article in the *Journal of Fundamental and Applied Sciences* [3] represents a logical continuation of the 2014 work, systematically investigating the effect of initial synthesis pH and crystallization temperature on the properties of MCM-bentonite. This is the most comprehensive optimization study on these materials.

The central issue is as follows: the molten bentonite supernatant contains a high concentration of Na^{+} ions (from NaOH used in fusion), which favors the formation of zeolite phases (thermodynamically more stable under alkaline conditions) at the expense of the mesoporous MCM-41 phase if the pH is not adjusted. The addition of acetic acid to lower the synthesis pH allows control of the competition between hexagonal mesophase nucleation and crystalline zeolite phases.

The study covered a pH range (8.5–11.5) at three temperatures (80, 100, and 130°C). The main conclusions are summarized in Table III.2.

Table III.2. Effect of synthesis pH and crystallization temperature on MCM-bentonite properties. Data from Ali-Dahmane et al. [3].

T (°C)	pH	Phase obtained	SBET (m^2/g)	Vmeso (cm^3/g)	Si/Al (cristalline)	Observation
80	8,5	Pure MCM-41	373	0,59	99	Best Al incorporation (amorphous part)

80	9,5	MCM-41 + lamellar phase.	—	—	—	Lamellar phase present
80	11,5	MCM-41 + lamellar phase.	—	—	—	Significant lamellar phase
100	8,5	Pure MCM-41	450	0,68	50	Good structural order
100	9,5	Pure MCM-41	494	0,72	43,64	Optimum: best order + texture
100	11,5	MCM-41 + lamellar phase.	—	—	—	Lamellar phase at high pH
130	8,5	Pure MCM-41	400	0,62	35	No lamellar phase
130	9,5	Pure MCM-41	420	0,63	32	Good materia
130	11,5	Pure MCM-41	417	0,60	30	No lamellar phase

The general rule established by this study is fundamental: as crystallization temperature decreases, synthesis pH must also be lowered to obtain pure hexagonal phase and suppress the lamellar phase. At 130°C, the hexagonal phase is stable across the entire pH range studied (8.5–11.5). This rule can be rationalized by temperature effects on aluminum incorporation: at lower temperatures, higher aluminum incorporation (which favors lamellar phases) requires lower pH to compensate.

III.2.3. Complete characterization of the optimized material (100°C, pH = 9.5)

The MCM-bentonite sample prepared at 100°C and pH = 9.5 was characterized using all available techniques. The nitrogen adsorption–desorption isotherm at 77 K is type IV (IUPAC), with a sharp inflection at $P/P_0 \approx 0.25$–0.35 corresponding to capillary condensation in mesopores (Dp ≈ 38 Å). An H4 hysteresis loop is observed at $P/P_0 > 0.45$, attributed to condensation in interparticle mesopores.

TEM images (Figure III.1) directly confirm the hexagonal channel structure by visualization of pore arrays in projection perpendicular to the channel axis. Direct measurement of periodicity gives a unit cell parameter of 42 Å, in very good agreement with XRD calculations (44 Å). SAED electron diffraction shows a well-ordered hexagonal pattern (Figure III.2).

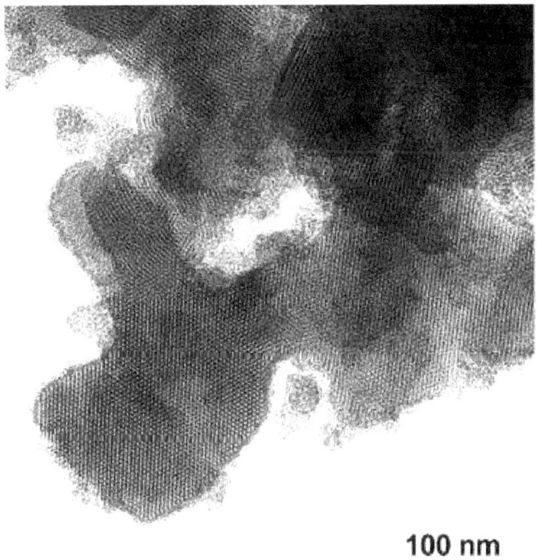

Figure III.1. TEM image of calcined MCM-bentonite (T = 100°C; pH = 9.5).

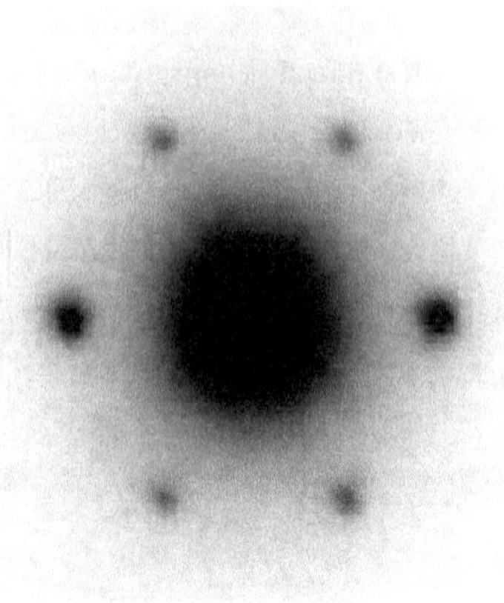

Figure III.2. Electron diffraction pattern of calcined MCM-bentonite (T = 100°C; pH = 9.5).

EDX point analysis distinguishes the Si/Al ratio in the crystalline hexagonal phase (Si/Al = 43.64) and in the amorphous phase (Si/Al = 12.43). This key result reveals that aluminum extracted from bentonite is distributed between both phases, but only aluminum incorporated into the hexagonal phase is directly catalytically active. This explains the lower Lewis acidity of Al-MCM-bentonite compared to standard Al-MCM (Si/Al = 13.47), despite their comparable overall Si/Al ratio.

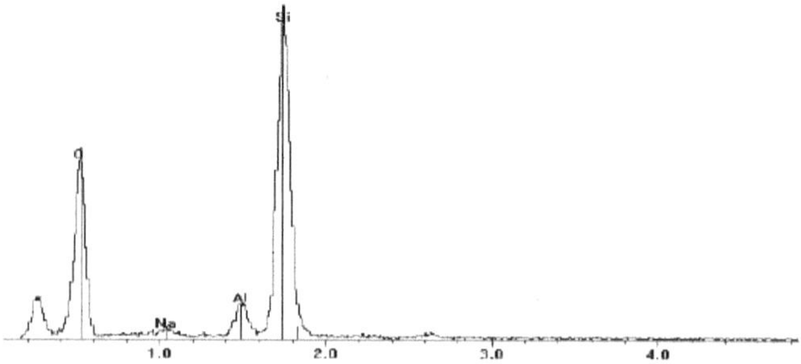

Figure III.3. EDX spectrum of the crystalline region of MCM-bentonite (T = 100°C; pH = 9.5).

III.3. References of Chapter III

1. Ali-Dahmane T., Brahmi L., Hamacha R., Bengueddach A. (2017). How to control the structural properties of purely siliceous MCM-41. *Annales de Chimie Sciences des Matériaux*, 40(3–4), 149–163.

2. Ali-Dahmane T., Adjdir M., Hamacha R., Villieras F., Bengueddach A., Weidler P.G. (2014). The synthesis of MCM-41 nanomaterial from Algerian bentonite: the effect of the mineral phase contents of clay on the structural properties of MCM-41. *Comptes Rendus Chimie*, 17(1–2), 1–6.

3. Ali-Dahmane T., Brahmi L., Hamacha R., Villieras F., Bengueddach A. (2017). Synthesis of MCM-41 nanomaterial from Algerian bentonite: influence of synthesis pH. *Journal of Fundamental and Applied Sciences*, 9(2), 636–649.

CHAPTER IV – Lewis Acidity of Al-MCM-41: Standard vs Bentonite Comparison

IV.1. Introduction and Problem Statement

The surface acidity of aluminosilicate materials is one of their most important properties for catalytic applications. In the case of Al-MCM-41, acidity results directly from the incorporation of aluminum into the siliceous framework, and its nature (Lewis vs Brønsted) as well as its intensity (site strength) determine catalytic selectivity and activity. The central question addressed in this chapter is the following: how does the Lewis acidity of Al-MCM-41 synthesized from Algerian bentonite compare with that of standard Al-MCM-41, and how can the observed differences be explained on rigorous structural and compositional bases?

This comparative study was published in *Bulletin of Chemical Reaction Engineering and Catalysis* in 2019 [1], and its results form the core of this chapter.

IV.2. Comparative Structural Characterization by XRD

A major observation from the XRD patterns of the three materials (Si-MCM-standard, Al-MCM-standard, and Al-MCM-bentonite) is as follows:

The isomorphic substitution Si → Al in Al-MCM-standard (Si/Al = 12.5) significantly degrades the hexagonal structural order compared to Si-MCM-standard (broader (100) peak and less resolved (110) and (200) reflections). In contrast, Al-MCM-bentonite exhibits a superior hexagonal organization compared to Al-MCM-standard, with well-resolved (110), (200), (210), and even (300) reflections.

This counterintuitive result can be explained by the different nature of aluminum in the two materials: in Al-MCM-standard, aluminum is homogeneously incorporated throughout the siliceous framework (including the pore walls), disturbing the uniformity of surfactant–silica interactions. In Al-MCM-bentonite, aluminum is mainly located in the interstitial amorphous phase, leaving the hexagonal phase more purely siliceous and therefore better organized (Figure IV.1).

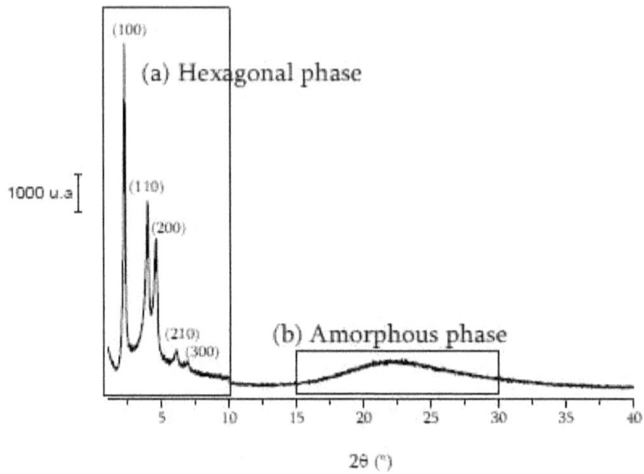

Figure IV.1. XRD pattern of calcined Al-MCM-bentonite: (a) low angle and (b) high angle [1].

The contraction of the unit cell upon calcination (Δa_0) confirms this behavior: Δa_0 = 0.9% for Al-MCM-bentonite compared to 12.0% for Al-MCM-standard. This low contraction indicates higher rigidity and cohesion of the pore walls in the bentonite-derived material, likely due to natural cations from bentonite reinforcing silicate chain linkages.

IV.3. Comparative Textural Properties

Table IV.1 compares the textural properties of the three reference materials studied in this chapter. The observed differences reflect the combined effect of aluminum content, source nature, and synthesis method on the microstructure.

Tableau IV.1. Comparison of textural properties of Si-MCM-standard, Al-MCM-standard, and Al-MCM-bentonite [1].

Property	Si-MCM-standard	Al-MCM-standard	Al-MCM-bentonite
a_0 calciné (Å)	41,60	43,94	44,51
Δa_0 (%)	9,8	12,0	0,9
SBET (m^2/g)	1154	983	674
Vmeso (cm^3/g)	0,84	0,76	0,42
Dp BJH (Å)	34,0	35,6	35,6
Wt (Å)	7,6	8,4	8,9
Si/Al EDX (hex.)	∞	13,47	43,64
Si/Al EDX (amorph.)	∞	—	12,43

The decrease in SBET and Vmeso from Si-MCM-standard to Al-MCM-standard and then to Al-MCM-bentonite reflects the effect of aluminum incorporation: the increase in pore diameter (Dp) and wall thickness (Wt) upon Al^{3+} introduction (larger ionic radius than Si^{4+}, 0.535 Å vs 0.260 Å) is accompanied by a reduction in accessible internal surface area.

IV.4. Lewis Acidity by FTIR-Pyridine

IV.4.1. FTIR spectra and identification of acid sites

The FTIR spectra of pyridine adsorption on Al-MCM-41 and MCM-bentonite after desorption at different temperatures are shown in Figure IV.2.

FTIR analysis of pyridine adsorption–desorption on Al-MCM-standard reveals intense and well-resolved bands characteristic of Lewis acid sites (1456 cm^{-1}, 1622 cm^{-1}) and Brønsted acid sites (1545 cm^{-1}, 1635 cm^{-1}), which remain significant even after desorption at 350°C, indicating strong and stable acidity.

For Al-MCM-bentonite, the characteristic acid bands are present but significantly weaker. New bands at 1445 cm^{-1} and 1596 cm^{-1}, attributed to hydrogen-bonded physisorbed pyridine on silanol groups (Py–H), appear and disappear completely after desorption at 250°C. These observations suggest that the surface of Al-MCM-bentonite is richer in non-condensed silanol groups, consistent with the higher contraction observed in Al-MCM-standard during calcination.

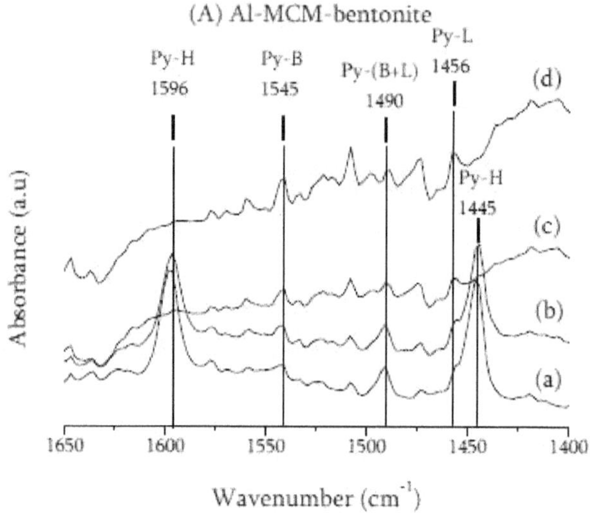

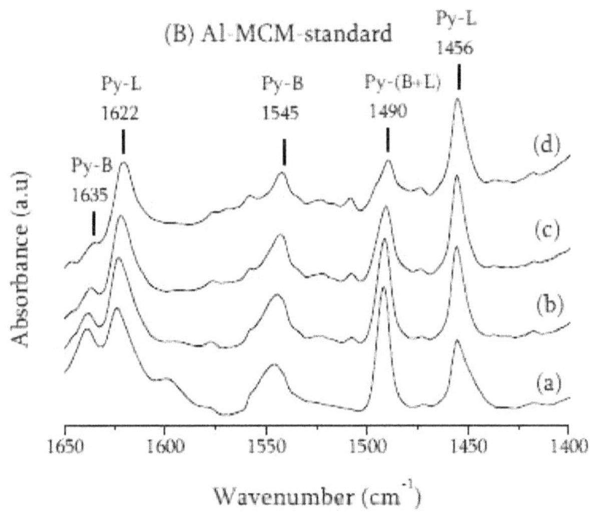

Figure IV.2. FTIR spectra of pyridine desorbed on calcined (A) Al-MCM-bentonite and (B) Al-MCM-standard under vacuum (3 Torr) at (a) 50°C, (b) 150°C, (c) 250°C, and (d) 350°C.

IV.4.2. Quantification of acid sites

Table IV.2 presents the concentrations of Lewis (CL) and Brønsted (CB) acid sites determined by integration of FTIR-pyridine bands at successive desorption temperatures.

Tableau IV.2. Acid site concentrations of Al-MCM-41 materials [1].

Acid property	**Catalyst**	**50°C**	**150°C**	**250°C**	**350°C**	**Total**
Total acidity (CL+CB) µmol/g	Al-MCM-standard	295.6	372.2	402.5	344.4	1414.8
	Al-MCM-bentonite	31.1	42.2	47.4	52.1	172.8
Brønsted CB (µmol/g)	Al-MCM-standard	92.1	124.8	142.8	85.9	445.6
	Al-MCM-bentonite	26.8	30.1	31.2	33.0	121.1
Lewis CL (µmol/g)	Al-MCM-standard	203.5	247.4	259.7	258.5	969.2
	Al-MCM-bentonite	4.3	12.1	16.2	19.1	51.7
CL/CB ratio	Al-MCM-standard	2.21	1.98	1.82	3.01	2.17
	Al-MCM-bentonite	0.16	0.40	0.51	0.57	0.42

These data reveal several important facts:

(i) Al-MCM-standard is approximately eight times more acidic than Al-MCM-bentonite in terms of total acidity.

(ii) In Al-MCM-standard, Lewis sites clearly dominate over Brønsted sites (CL/CB = 2.17), consistent with efficient isomorphic incorporation of tetrahedrally coordinated Al.

(iii) In Al-MCM-bentonite, Brønsted sites are relatively more significant (CL/CB = 0.42), reflecting the presence of partially extra-framework amorphous aluminum.

The thermal stability of acid sites also differs: in Al-MCM-standard, Lewis acidity remains nearly constant from 50 to 350°C (203.5 → 258.5 μmol/g), indicating strong and thermally stable sites. In Al-MCM-bentonite, CL increases from 4.3 to 19.1 μmol/g, indicating moderate-strength Lewis sites and progressive desorption of weakly bound pyridine.

IV.5. Correlation between Acidity and Aluminum Content

The correlation between aluminum content and Lewis acidity is analyzed using Si/Al ratios determined by EDX. For Al-MCM-standard, Si/Al = 13.47 in the hexagonal phase, leading to CL = 969 μmol/g. For Al-MCM-bentonite, Si/Al = 43.64, leading to CL = 51.7 μmol/g. The ratio CL(standard)/CL(bentonite) ≈ 18.7 is significantly higher than the simple Si/Al ratio (≈ 3.2), suggesting that aluminum incorporation in Al-MCM-bentonite is less catalytically effective, likely due to the inaccessibility of aluminum atoms located in the amorphous phase.

This is consistent with the work of Kumar et al. [2], which showed that aluminum derived from natural sources is not always catalytically active, with a significant fraction existing in separate alumina-like phases inaccessible to reactants.

IV.6. Allylation Reaction: Proof of Lewis Acidity Concept

The allylation reaction of benzaldehyde with allyltrimethylsilane (TMS-allyl) was used as a test reaction to evaluate the Lewis acidity of both materials (Figure IV.3).

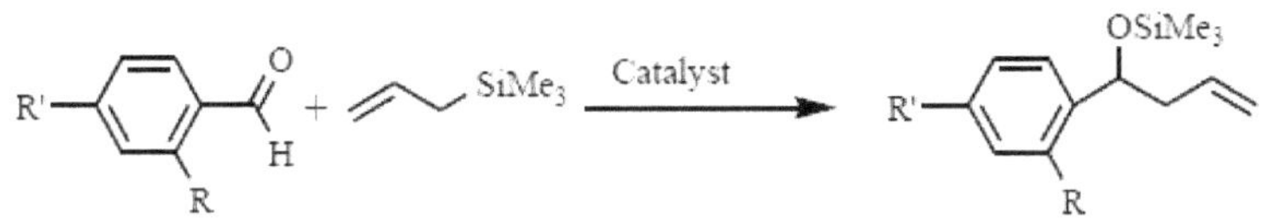

Figure IV.3. Allylation of aromatic aldehydes with allyltrimethylsilane.

A plausible mechanism is proposed involving the role of Lewis acid sites of Al-MCM-41 in activating substituted benzaldehydes in the presence of allyltrimethylsilane (Figure IV.4) [3].

Upon adsorption of benzaldehyde on the catalyst, strong chelation occurs between the carbonyl oxygen and the Lewis acid site of Al-MCM-41 (Scheme 1, A), facilitating the addition of allyltrimethylsilane and formation of a C–C bond (Scheme 1, B). Electron rearrangement in the intermediate (Scheme 1, C) leads to formation of the allylic product (Scheme 1, D), regenerating the Lewis acid site for a new catalytic cycle.

The transition from (C) to (D) shows that a second allyltrimethylsilane attack is not possible in the presence of Al-MCM-41, which explains why diallylated products are not observed.

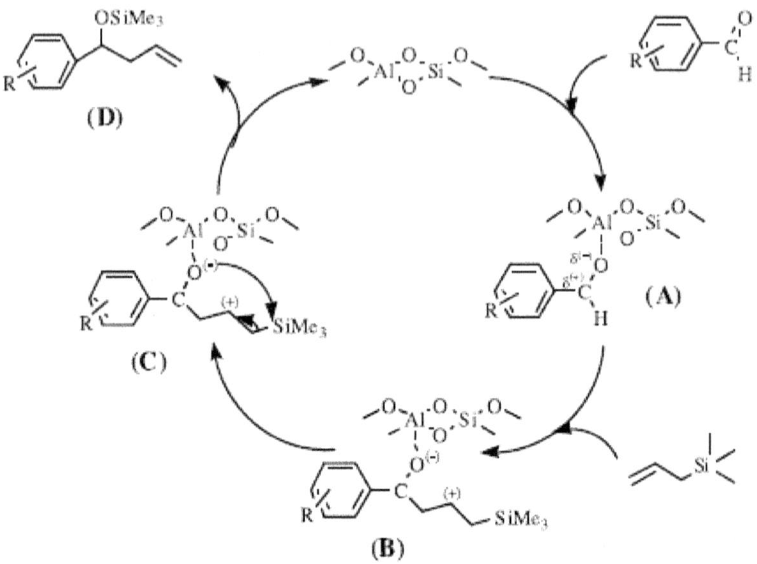

Figure IV.4. Proposed reaction mechanism showing the role of Al-MCM-41 as a Lewis acid in allylation reactions [3].

The results (Table IV.3) confirm a direct correlation between Lewis acidity measured by FTIR-pyridine and catalytic activity: Al-MCM-standard (strong Lewis acidity) gives 99% yield in 4 h, while Al-MCM-bentonite (moderate Lewis acidity) does not exceed 20% yield even after 24 h.

Tableau IV.3. Comparative results of catalytic allylation reactions [1].

Substrate	Catalyst	T (°C)	TMS-allyl (equiv.)	Time	Yield (%)
Benzaldehyde	Al-MCM-standard	35	3	4 h	99
Benzaldehyde	Al-MCM-bentonite	35	3	24 h	20
4,4'-dichlorobenzaldehyde	Al-MCM-standard	35	3	45 min	100

4,4'-dichlorobenzaldehyde	Al-MCM-bentonite	35	3	20 h	36
4-nitrobenzaldehyde	Al-MCM-standard	35	1.5	30 min	99
4-nitrobenzaldehyde	Al-MCM-bentonite	35	1.5	20 h	17
Benzaldehyde	Al-MCM-standard	0	3	24 h	traces
Benzaldehyde	Al-MCM-bentonite	0	3	24 h	traces

Increasing the amount of Al-MCM-bentonite up to 100 mol% does not significantly improve the yield (≤ 20%, Figure IV.5), confirming that catalytic performance is limited by the quality (Lewis strength) of the sites rather than their quantity. This is consistent with the lower Lewis acidity of Al-MCM-bentonite, insufficient to efficiently activate aldehydes in this reaction, which requires moderate to strong Lewis acidity.

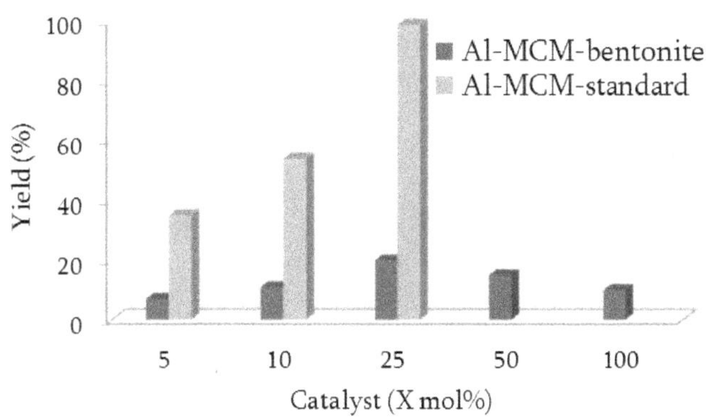

Figure IV.5. Effect of the amount of Al-MCM-bentonite or Al-MCM-standard on the yield of the allylation reaction of benzaldehyde with allyltri-methylsilane [1].

IV.7. References of Chapter IV

1. Ali-Dahmane T., Brahmi L., Hamacha R., Hacini S., Bengueddach A. (2019). Comparison of Lewis acidity between Al-MCM-41 pure chemicals and Al-MCM-41 synthesized from bentonite. *Bulletin of Chemical Reaction Engineering and Catalysis*, 14(2), 358–368.

2. Kumar P., Mal N., Oumi Y., Yamana K., Sano T. (2001). Mesoporous materials prepared using coal fly ash as the silicon and aluminium source. *Journal of Materials Chemistry*, 11(12), 3285–3290.

3. Brahmi L., Ali-Dahmane T., Hamacha R., Hacini S. (2016). Catalytic performance of Al-MCM-41 catalyst for the allylation of aromatic aldehydes with allyltrimethylsilane: comparison with $TiCl_4$ as Lewis acid. *Journal of Molecular Catalysis A: Chemical*, 423, 31–40.

CHAPTER V – Green Heterogeneous Catalysis: Eco-compatible Synthesis of Formamidine Derivatives

V.1. Introduction: Importance of Formamidines in Medicinal Chemistry

Formamidines, compounds with the general formula R–N=CH–NR'$_2$, constitute an important functional class in organic chemistry, with major applications in medicinal chemistry, agrochemistry, and materials synthesis. Their reactivity versatility arises from their dual character: nucleophilic (at the nitrogen bearing substituent R) and electrophilic (at the carbon of the C=N group), making them reactive toward both nucleophilic and electrophilic agents [1,2].

In medicinal chemistry, formamidine derivatives are valuable precursors for accessing numerous nitrogen-containing heterocycles with pharmacological activity. Lüth and Löwe [3] demonstrated their utility in the synthesis of antiparasitic diaminopyrimidines (antileishmanial agents). Zhang et al. [4] prepared formamidines with antibacterial activity. In our research group, formamidines have been used as intermediates for the synthesis of pyridines [5], pyrimidines [6], and imidazoles via multicomponent cyclization strategies.

Formamidines are also used as selective protecting groups for primary amines in organic synthesis, as demonstrated by Toste et al. [7] and Vincent et al. [8]. Their deprotection can be carried out under mild conditions (light acidic hydrolysis), making formamidines orthogonal protecting groups.

V.2. State of the Art in Formamidine Synthesis Methods

Several synthetic methods for formamidine derivatives have been reported in the literature [9–14]:

Method 1 — Amine/DMF-DMA condensation (catalyst-free)

The most widely used method is the direct condensation of primary amines with DMF-DMA (N,N-dimethylformamide dimethyl acetal). Under classical thermal conditions (reflux in aromatic solvents such as p-xylene or toluene, 120–140°C, 5–16 h), satisfactory yields (75–90%) are obtained. However, long reaction times and high temperatures remain major limitations.

Method 2 — Homogeneous acid catalysis

The use of homogeneous acid catalysts (HCl, AcOH, p-TsOH) accelerates the reaction but leads to selectivity issues and difficulties in catalyst separation.

Method 3 — Microwave activation

Microwave irradiation significantly reduces reaction times (10–30 min), but requires specialized equipment and does not improve the green chemistry profile in terms of solvent use.

The original contribution of Datoussaid, Ali-Dahmane et al. [15] consists of using Al-MCM-bentonite as a solid heterogeneous acid catalyst for formamidine synthesis under solvent-free conditions at room temperature, representing a significant advance in efficiency, reaction time, and green chemistry compatibility.

V.3. Optimization of Catalytic Conditions

V.3.1. Model reaction and parametric study

The model reaction selected for optimization is the condensation of 3-aminothiophene-2-carbonitrile with DMF-DMA to form N'-(3-cyano-4,5,6,7-tetrahydrobenzo[b]thiophen-2-yl)-N,N-dimethylformamidine:

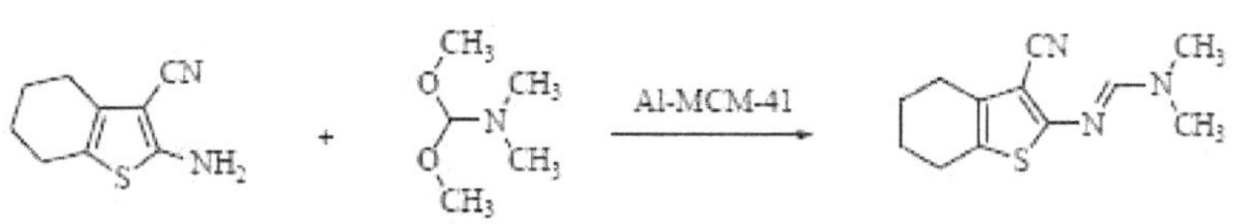

This reaction is well documented in the literature under thermal conditions [14], with a reference yield of 78% after 5 h reflux in p-xylene (135°C), and thus serves as a comparative benchmark.

The parametric study examined the effect of: (i) solvent nature; (ii) catalyst loading (2–30 wt%); (iii) reaction temperature; and (iv) reaction time. The results are summarized in Table V.1.

Tableau V.1. Optimization of formamidine synthesis using Al-MCM-41 catalysis [15].

Entry	Solvent	Temp.	Cat. (wt%)	Time	Yield (%)	Observation
1	p-xylene	Room temp.	0	48 h	0	No reaction
2	p-xylene	Reflux (135°C)	0	5 h	78	Thermal method [14]
3	DMF	Reflux	0	16 h	19	Poor
4	Solvent-free	Room temp.	30	1 h	95	Excess catalyst
5	Solvent-free	Room temp.	10	1 h	95	Optimal
6	Solvent-free	Room temp.	5	8 h	58	Slow
7	Solvent-free	Room temp.	2	16 h	10	Very slow
8	p-xylene	Room temp.	10	1 h	95	Solvent unnecessary
9	DMF	Room temp.	10	1 h	95	Solvent unnecessary
10	Toluene	Room temp.	10	1 h	95	Solvent unnecessary
11	$CHCl_3$	Room temp.	10	1 h	95	Solvent unnecessary

The results lead to several important conclusions:

- The reaction does not occur at room temperature without catalyst, even after 48 h.
- An optimal catalyst loading of 10 wt% is identified; below this value conversion decreases, above it no improvement is observed.
- The solvent is completely unnecessary: identical yields (95%) are obtained with or without solvent.
- Optimal conditions are therefore: 10 wt% Al-MCM-41, solvent-free, room temperature, 1 h. These conditions are significantly superior to the thermal method (78%, 5 h, reflux).

V.3.2. Mechanism of activation by Al-MCM-41

The proposed mechanism involves activation of both reactive functions of DMF-DMA by Lewis and Brønsted acid sites of Al-MCM-41 (Figure V.1):

Step 1 — Activation of DMF-DMA:

One methoxy group is activated by coordination or protonation at an acid site, leading to methanol departure and formation of an iminium intermediate (Me_2N^+=CH–OMe).

Step 2 — Nucleophilic addition of amine:

The primary amine (R–NH_2) attacks the electrophilic carbon of the iminium, forming a zwitterionic intermediate (Me_2N–CH(OMe)–NHR).

Step 3 — Elimination of second methanol:

Activation of the second methoxy group leads to its elimination and formation of the final formamidine product (Me_2N–CH=N–R), regenerating the catalyst.

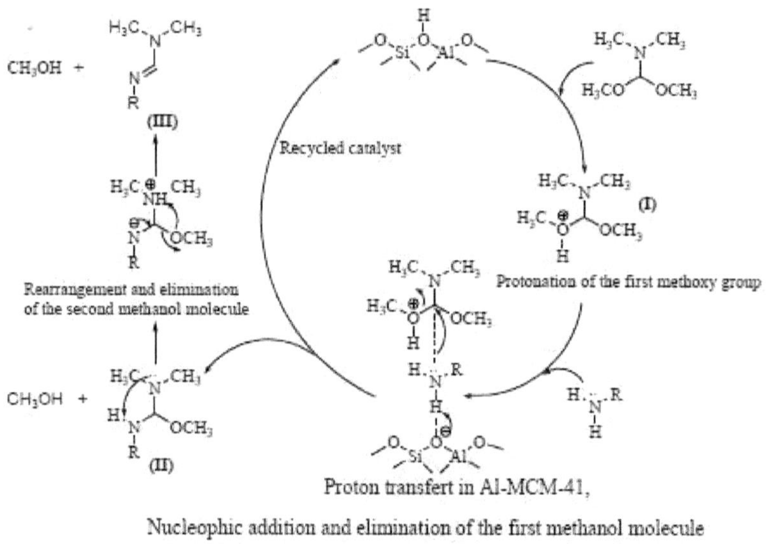

Figure V.1. Proposed catalytic cycle for the synthesis of formamidine [15].

V.4. Reaction scope: formamidine library

The optimized conditions (10 wt% Al-MCM-41, solvent-free, room temperature, 30 min–1 h) were applied to eight structurally diverse primary amines (aromatic and heteroaromatic). Results are summarized in Table V.2.

Tableau V.2 Synthesis of formamidines 2a–2h [15]

Product	Starting amine	R.cat (%)	R.therm (%)	t(cat)	t(therm)	Improvement
2a	3-Amino-4-Ph-6-p-tolylpyridine	95	75	30 min	3h30	7.0×
2b	3-Amino-4,6-diphenylpyridine	97	87	30 min	4h30	9.0×
2c	3-Amino-4-(4-ClPh)-6-p-tolylpyridine	97	90	30 min	4h30	9.0×
2d	3-Amino-6-(4-BrPh)-4-(4-ClPh)	98	79	45 min	3h45	5.0×
2e	3-Amino-4-(4-FPh)-6-(3-MeOPh)	93	88	30 min	3h30	7.0×
2f	3-Amino-6-(4-FPh)-4-(thiophen-2-yl)	96	82	30 min	4h30	9.0×
2g	Methyl 2-amino-THBT-3-carboxylate	99	78	1 h	12 h	12.0×
2h	3-Amino-THBT	96	78	1 h	5 h	5.0×

The results demonstrate:

- Catalytic method always outperforms thermal method in yield and time.
- Reaction time is reduced by a factor of 5–12.
- Substituent effects have little influence on catalytic yield, showing robustness.
- Compounds 2a, 2e, 2f, and 2g are new.

V.5. Catalyst recyclability and stability

V.5.1. Six-cycle recyclability tests

Catalyst recyclability was evaluated over six consecutive cycles using model reaction 2h (Table V.3 and Figure V.2). After each cycle, the catalyst was recovered by filtration, washed with absolute ethanol (3 × 10 mL), and dried at 100°C for 3 h. No calcination was required between cycles.

Tableau V.3. Recyclability of Al-MCM-41 catalyst [15]

Cycle	1	2	3	4	5	6
Yield (%)	98	96	93	91	87	80
Loss (%)	0	0	3.1	5.2	9.4	16.7

The recyclability of Al-MCM-41 significantly reduces catalytic cost. Assuming a production cost of 620 DA/g, the catalyst cost per mmol of product decreases from 0.62 DA/mmol (cycle 1) to 0.10 DA/mmol (cycle 6), corresponding to an ~84% reduction.

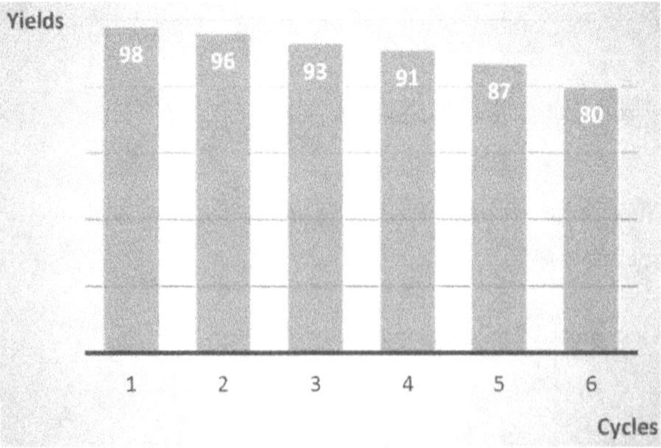

Figure V.2. Recyclability of catalyst.

V.5.2. Environmental and economic advantage

The recyclability of Al-MCM-41 significantly reduces catalytic cost. Assuming a production cost of 620 DA/g, the catalyst cost per mmol of product decreases from 0.62 DA/mmol (cycle 1) to 0.10 DA/mmol (cycle 6), corresponding to an ~84% reduction.

V.6. Characterization of synthesized formamidines

All products were purified by recrystallization (ethanol washing) without column chromatography. They were characterized by:

- ^1H NMR (400 MHz, $CDCl_3$)
- ^{13}C NMR (100 MHz, $CDCl_3$)
- FTIR (KBr pellets)
- Melting point

The characteristic ^1H NMR signal is a singlet at 8.62–8.84 ppm corresponding to the N=CH proton. In ^{13}C NMR, the N=CH carbon appears at $\delta \approx$ 164–165 ppm. In IR, ν(CN) appears at 2195–2214 cm^{-1} and ν(C=N) at 1620–1650 cm^{-1}.

V.7. References of Chapter V

1. Erian A.W., Sherif S.M., Alassar A.-Z.A., Elkholy Y.M. (1994). Reactions of formamidines with different electrophilic systems. *Tetrahedron*, 50(7), 1877–1884.

2. Meyers A., Marra J.M. (1985). The use of formamidines as protecting groups for primary amines. *Tetrahedron Letters*, 26(48), 5863–5866.

3. Lüth A., Löwe W. (2008). Design and synthesis of novel diaminopyrimidines as antileishmanial agents. *European Journal of Medicinal Chemistry*, 43(7), 1478–1488.

4. Zhang M., Jiang L., Tao J., Pan Z., He M., Su D., He G., Jiang Q. (2019). Synthesis and antibacterial activity of novel formamidine-based compounds. *Bioorganic and Medicinal Chemistry*, 27(11), 2268–2279.

5. Belhadj F., Kibou Z., Cheikh N., Choukchou-Braham N., Villemin D. (2015). Synthesis of cyanopyridines via multicomponent condensation reactions. *Tetrahedron Letters*, 56(42), 5999–6002.

6. Medjdoub A., Belhadj F., Merzouk A.S., Hamed Y.B., Kibou Z., Choukchou-Braham N., Merzouk H. (2020). Synthesis of pyrimidines bearing 2-cyanopyridine moiety. *Chemistry Papers*, 74(3), 903–913.
7. Toste D., McNulty J., Still I.W. (1994). Use of formamidines as protecting groups for primary amines. *Synthetic Communications*, 24(12), 1617–1624.

8. Vincent S., Mons S., Lebeau L., Mioskowski C. (1997). N,N-Dimethylformamide dimethyl acetal: activation reagent for amine protection. *Tetrahedron Letters*, 38(44), 7527–7530.

9. Meesala R., Arshad A.S.M., Mordi M.N., Mansor S.M. (2014). Synthesis of N,N-dimethylformamidines. *Tetrahedron Letters*, 55(50), 7023–7025.

10. Dar B.A., Ahmad S.N., Wagay M.A., Hussain A., Ahmad N., Bhat K.A., Khuroo M.A., Sharma M., Singh B. (2013). Synthesis of formamidines using Cs_2CO_3. *Tetrahedron Letters*, 54(38), 4880–4884.

11. Han Y., Cai L. (1997). Efficient synthesis of N-arylformamidines. *Tetrahedron Letters*, 38(29), 5423–5426.

12. Jalani H.B., Pandya A.N., Pandya D.H., Sharma J.A., Sudarsanam V., Vasu K.K. (2012). Synthesis of N,N-disubstituted formamidines. *Tetrahedron Letters*, 53(32), 4062–4064.

13. Abu-Shanab F.A., Sherif S.M., Mousa S.A. (2009). Chemistry of formamidines and related compounds. *Journal of Heterocyclic Chemistry*, 46(5), 801–827.

14. Patil P.S., Chavan S.M., Toche R.B., Bhole A.D., Patil V.M., Aware P.B. (2017). Synthesis of formamidine derivatives via thermal condensation. *IOSR Journal of Applied Chemistry*, 10(11), 96–100.

15. Datoussaid Y., Ali-Dahmane T., Mehiaoui N., Brahmi L., Kibou Z., Hamacha R., Choukchou-Braham N. (2021). Ecofriendly synthesis of formamidine derivatives catalyzed by Al-MCM-41 mesoporous materials. *Letters in Organic Chemistry*, 18, 1–6.

CHAPTER VI – Water Treatment by Adsorption on MCM-41 Derived from Maghnia Bentonite

VI.1. Problem of water pollution in Algeria

Wastewater management is one of the most pressing environmental challenges facing Algeria. The rapid industrial development over recent decades, particularly in the textile, chemical, agro-food, petroleum, and mining sectors, generates large volumes of aqueous effluents containing organic and inorganic pollutants. The discharge of these effluents into hydrographic networks without adequate treatment severely threatens water resource quality and aquatic ecosystems.

Among the most widespread pollutants, textile dyes are particularly concerning: they account for approximately 10–15% of total dye production and are difficult to remove using conventional treatment processes (coagulation–flocculation, aerobic biodegradation) due to their high chemical and photochemical stability. Methylene blue (MB, cationic dye, MW = 319.85 g/mol, λmax = 664 nm) is the most commonly used model dye in adsorption studies because of its well-documented characteristics.

In this context, the development of efficient, low-cost adsorbents produced from local Algerian resources represents a strategic issue in national environmental research. MCM-41 synthesized from Maghnia bentonite, with moderate production cost, high specific surface area (494 m^2/g), and uniform mesoporous channels, constitutes a relevant solution to this challenge [1].

VI.2. Characterization of MCM-bentonite as an adsorbent

VI.2.1. X-ray diffraction analysis

The diffractogram in Figure VI.1 shows that the natural bentonite is dominated by clay minerals, primarily montmorillonite, illite, and kaolinite, and reveals the presence of quartz, a non-clay constituent.

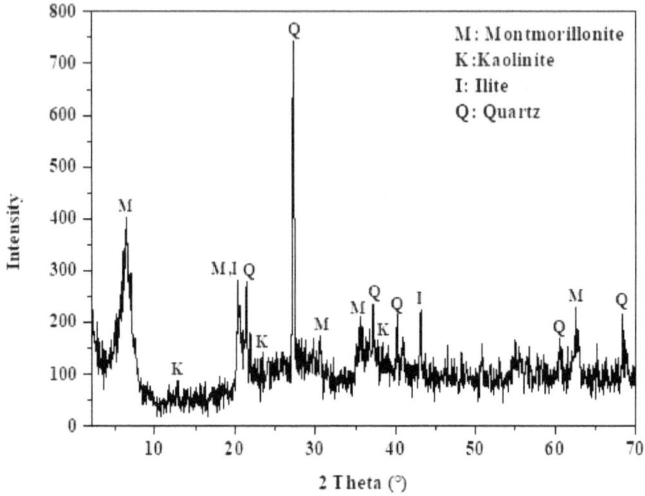

Figure VI.1. XRD patterns of the bentonite.

The diffractograms in Figures VI.2a and 2b show the complete disappearance of the characteristic mineral phases of bentonite, confirming that the alkaline fusion process primarily enabled the extraction of silica, subsequently used for the synthesis of MCM-41. Figures 2a$_1$ and 2b$_1$ present a detailed view of the diffractograms in the low-angle region (2θ = 2-10°), allowing a closer observation of the mesostructure.

According to the literature [2, 3], the mesoporous material of the MCM-41 type is characterized by a strong peak corresponding to the (100) reflection, indicative of a well-ordered hexagonal structure. The appearance of less intense secondary peaks, corresponding to the (110), (200), and (210) reflections, attests to a well-organized mesostructured phase. Moreover, the intensity and width of these peaks indicate the degree of structural order within the hexagonal channels of MCM-41.

Figure VI.2 thus presents the diffractograms of MCM-bentonite before (2a, 2a$_1$) and after calcination (2b, 2b$_1$). The as-synthesized material shows only the (100) reflection peak. After calcination, aimed at removing the structuring agent and thereby freeing the pores, the intensity of this peak significantly increases (I_{100} (as-synthesized) = 1303 a.u. vs. I_{100} (calcined) = 10017 a.u.). Additionally, the emergence of three further peaks corresponding to the (110), (200), and (210) reflections confirms that the material prepared from bentonite as a silica source exhibits a well-ordered hexagonal channel organization. The lattice parameters, a_0 = 41.77 Å (as-synthesized) and 45.79 Å (calcined), are consistent with those reported in the literature [2, 3].

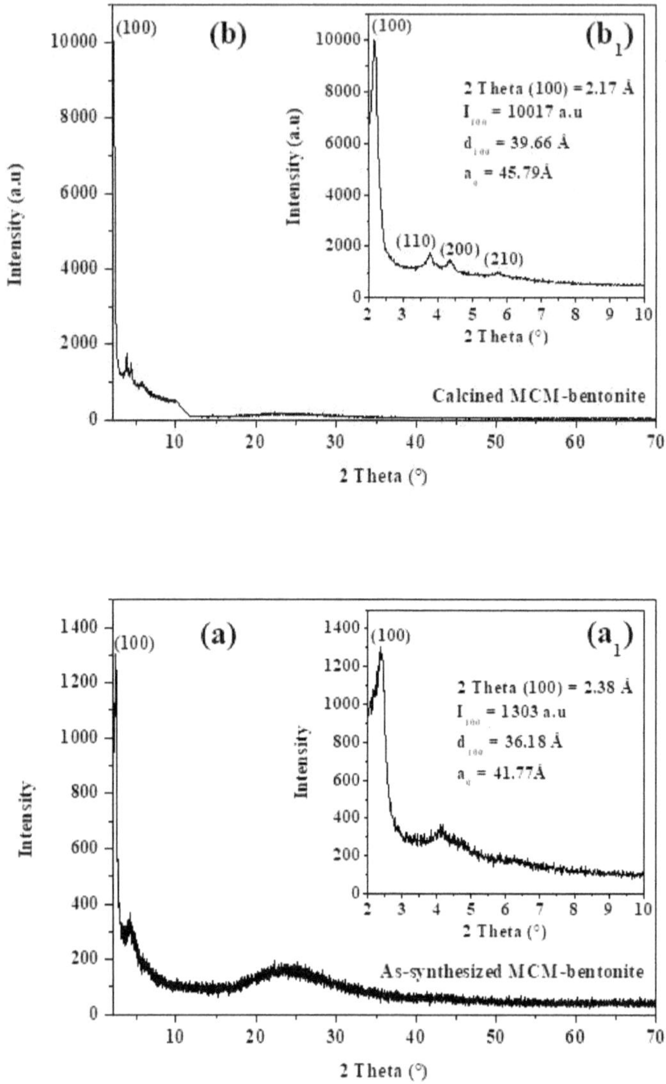

Figure VI.2. XRD patterns of MCM-bentonite before (a, a_1) and after calcination (b, b_1). Wide-angle range: 2θ = 2-70° (a, b); low-angle range: 2θ = 2-10° (a_1, b_1).

VI.2.2. Infrared spectroscopy

The FTIR spectra of MCM-41 synthesized from bentonite as a silica source are presented in Figure VI.3. These spectra exhibit the main bands reported in the literature [4, 5], which are characteristic of mesoporous materials of the MCM-41 type.

A broad band observed between 3300 and 3600 cm^{-1} is attributed to O–H stretching vibrations (ν(O-H)) arising from adsorbed water and surface silanol groups (Si-OH).

The bands at 2921 and 2851 cm^{-1} correspond to C-H stretching vibrations (ν(C-H)) of the $-CH_2-$ and $-CH_3$ groups originating from the structuring agent CTAB.

The region between 1620 and 1660 cm^{-1} is also associated with O-H bending vibrations related to adsorbed water.

The band at 1500 cm^{-1} is assigned to C-H bending vibrations (δ(C-H)) of the $-CH_2-$ and $-CH_3$ groups of CTAB.

Furthermore, the bands observed between 950 and 1300 cm^{-1} correspond to the asymmetric stretching vibrations of Si-O-Si bonds, while the band at 810 cm^{-1} is attributed to the symmetric stretching of Si-O-Si, and the band at 465 cm^{-1} to O-Si-O bending vibrations.

After calcination, several bands associated with the organic groups of the structuring agent disappear. Specifically, the bands at 2923 and 2853 cm^{-1}, corresponding to the C-H stretching vibrations of $-CH_2-$ and $-CH_3$ groups, as well as the band at 1479 cm^{-1}, assigned to C-H bending vibrations, are no longer observed. The disappearance of these bands confirms the complete decomposition and removal of the structuring agent, thereby enabling the liberation of the pores in the MCM-bentonite.

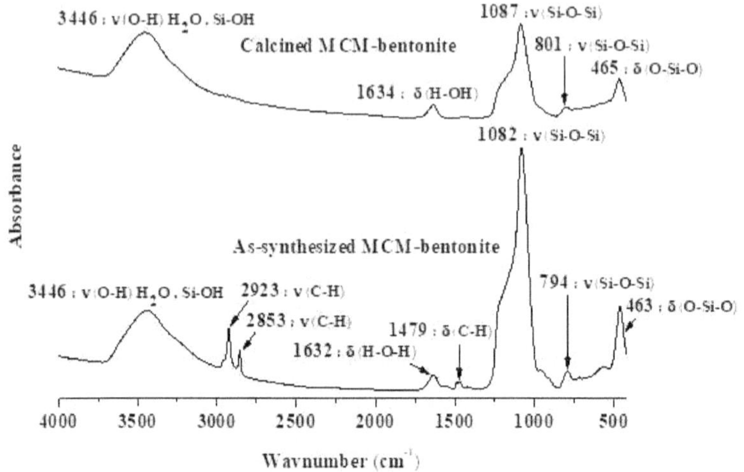

Figure VI.3. FTIR spectra of as-synthesized and calcined MCM-bentonite.

VI.2.3. Nitogen adsorption-desorption at 77 K analysis

Figure VI. 4a presents the N_2 adsorption-desorption isotherm of the calcined MCM-bentonite sample. The isotherm exhibits the characteristics of a type IV curve according to the IUPAC classification [6] with a distinct capillary condensation step occurring at relative pressures between 0.25 and 0.36. This behavior is typical of mesoporous materials with a narrow and uniform pore size distribution. The position of the inflection point reflects the average mesopore diameter, while the sharpness of the step indicates the homogeneity of the pores, as highlighted in Figure VI. 4b.

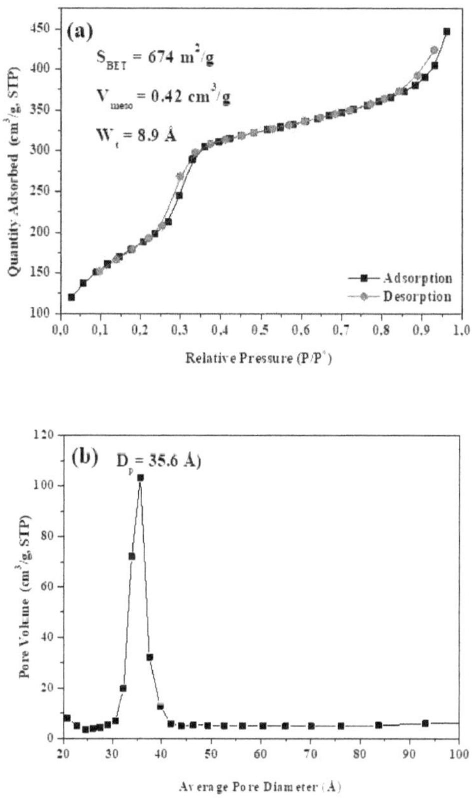

Figure VI. 4. (a) Nitrogen adsorption-desorption isotherm and (b) pore size distribution derived from the N_2 adsorption isotherm of calcined MCM-bentonite.

The textural properties of the material were evaluated (Figure VI. 4a), revealing a

:

- Specific surface area (S_{BET}) of 674 m^2/g.

- A mesopore volume (V_{meso}) of 0.42 cm^3/g.

- The average pore diameter (D_p), determined using the BJH method, was 35.6 Å (Figure VI. 4b).

- Using the lattice parameter (a_0) obtained from XRD measurements, the wall thickness (W_t) of the mesoporous channels was calculated according to $Wt = a_0 - D_p$, resulting in a value of 8.9 Å.

These findings demonstrate the successful formation of well-ordered mesoporous channels within the MCM-bentonite structure.

VI.3. Adsorption studies

VI.3.1. Effect of contact time on adsorption

The effect of contact time on the adsorption of methylene blue (MB) onto MCM-bentonite was investigated for an initial concentration of 10 ppm in a 20 mL solution, using 5 mg of adsorbent at ambient initial pH. The contact time was varied from 5 to 240 minutes, and the results are shown in Figure VI.5.

A rapid adsorption was observed during the first five minutes, with an adsorbed amount q_t = 30.89 mg/g and a removal efficiency of R = 77.5 %. This indicates that, at the initial stage, the active sites of MCM-41 are readily accessible, promoting a fast adsorption of MB. This initial rapid phase is particularly advantageous for practical water treatment applications targeting efficient MB removal. After 45 minutes, the maximum adsorption capacity was reached (q_t = 38.695, R = 98.94%), suggesting that most of the active sites on MCM-41 are saturated at this stage.

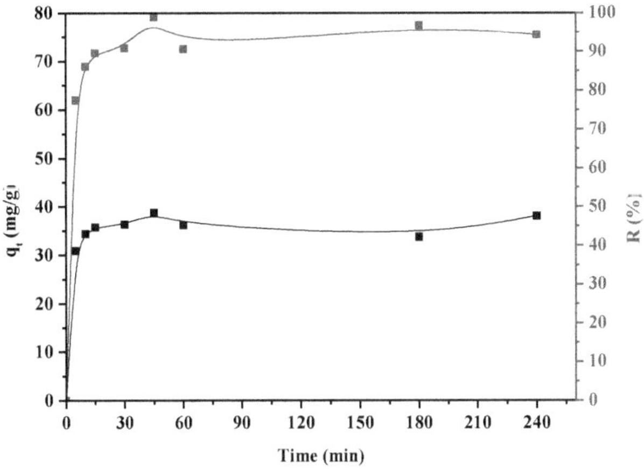

Figure VI.5. Effect of contact time on the removal efficiency and adsorption capacity of MB by MCM-bentonite: (MB concentration = 10 ppm, volume of the solution = 20 mL, MCM-bentonite dosage = 5 mg, pH = 7.5, room temperature = 17 °C).

The kinetic behavior of MB adsorption onto mesoporous MCM-bentonite was examined through two established models: pseudo-first-order and pseudo-second-order.

The linear form of the pseudo-first-order model was originally proposed by Lagergren (1898) [7], as indicated in equation (VI.1), while Ho and McKay (1999) [8] later introduced the linear representation of the pseudo-second-order model, given in equation (VI.2).

$$\ln(q_e - q_t) = lnq_q - k_1 t \qquad (VI.1)$$

$$\frac{t}{q_t} = \frac{1}{k_2 q_e^2} + \frac{t}{q_e} \qquad (VI.2)$$

Where q_e and q_t (mg/g) are the adsorption capacities at equilibrium and at time t, respectively, and k_1 (min^{-1}) and k_2 (g/mg.min) are the rate constants for the pseudo-first-order and pseudo-second-order models.

For the pseudo-first-order model, ln(q_e-q_t) was plotted versus time (Figure IV.6(a)), while for the pseudo-second-order model, t/q was plotted versus time (Figure VI.6(b)). The kinetic parameters, derived from the slopes and intercepts of the linearized plots, are summarized in Table VI.1.

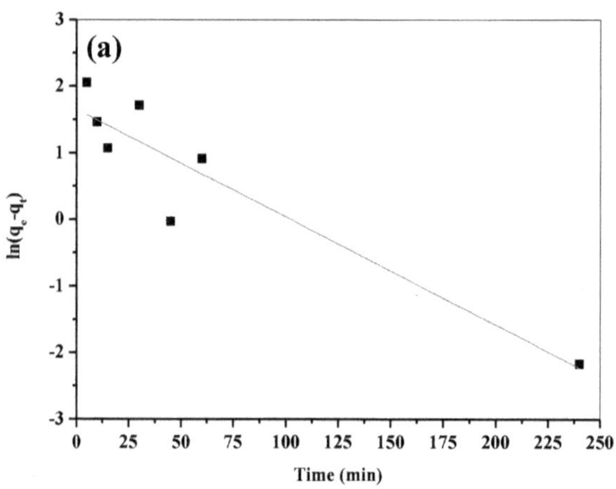

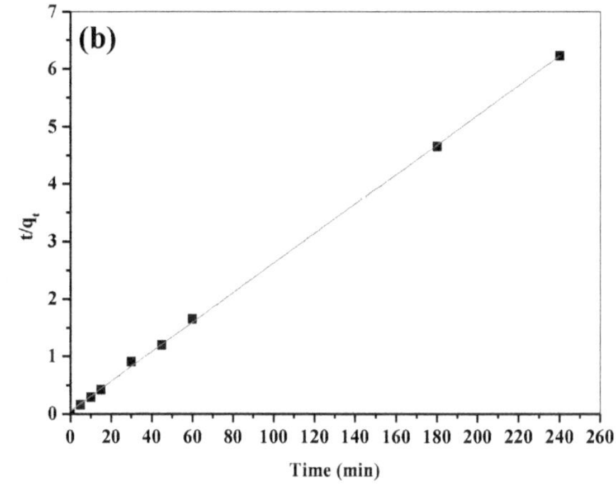

Figure VI.6. Kinetic models for MB adsorption onto MCM-bentonite: (a) pseudo-first-order model and (b) pseudo-second-order model.

According to Table VI.1, the pseudo-second-order model provides an excellent fit to the adsorption data, with a very high correlation coefficient ($R^2 > 0.999$) compared to the pseudo-first-order model ($R^2 > 0.842$). The calculated adsorption capacity ($q_{e,cal}$ = 38.895 mg/g) is in close agreement with the experimental value ($q_{e,exp}$ = 38.695 mg/g), confirming that the pseudo-second-order model reliably describes the process and that a heterogeneous sorption mechanism is likely responsible for MB adsorption onto MCM-bentonite [9].

Table VI.1. Kinetic parameters for the adsorption of MB onto MCM-bentonite.

	Pseudo-first-order			Pseudo-second order		
$q_{e,exp}$ (mg/g)	k_1 (min^{-1})	$q_{e,cal}$ (mg/g)	R^2	k_2 (g.min/mg)	$q_{e,cal}$ (mg/g)	R^2
38.695	0.0368	5.222	0.842	0.0130	38.895	0.999

VI.3.2. Effect of adsorbent dosage on adsorption

The effect of adsorbent mass on MB removal was investigated using a 10 ppm solution (20 mL) with a contact time of 45 minutes at natural pH (7.5). The adsorbent mass was varied from 2 to 200 mg, and the results are shown in Figure VI.7.

It is evident that increasing the adsorbent dose from 2 to 5 mg significantly enhances MB removal, reaching a maximum efficiency of 98.94%. This improvement is attributed to the higher number of active sites available on the adsorbent.

However, the specific adsorption capacity decreases as the adsorbent mass continues to rise. Although adding MCM-bentonite increases the total number of active sites, the MB concentration remains fixed at 10 ppm, leaving a substantial fraction of sites unoccupied.

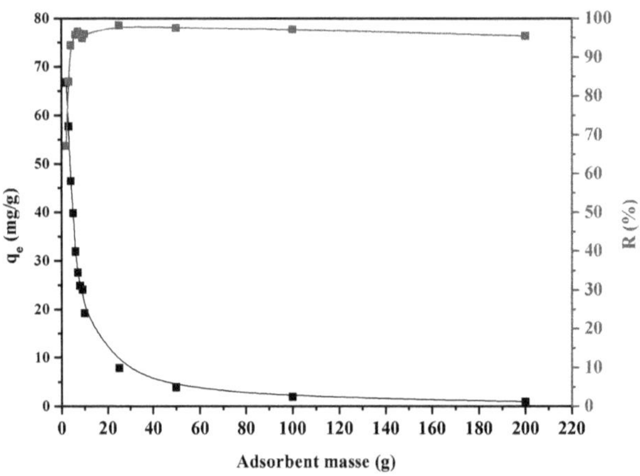

Figure VI.7. Effect of adsorbent mass on the removal efficiency and adsorption capacity of MB by MCM-bentonite: (MB concentration = 10 ppm, volume of the solution = 20 mL, contact time = 45 min, initial pH = 7.5, room temperature = 17 °C).

VI.3.3. Effect of initial methylene blue concentration

The study of the effect of the initial MB concentration on the adsorption process using MCM-bentonite as an adsorbent is essential for determining the adsorption isotherm. Experiments were conducted at a natural pH of 7.51, with an adsorbent mass of 5 mg, at room temperature.

As shown in Figure VI.8, the adsorption capacity (q_e) increases with the rise in MB concentration from 5 to 50 mg/L. A higher adsorbate concentration promotes the approach and interaction of MB molecules with the available adsorption sites on MCM-bentonite [10].

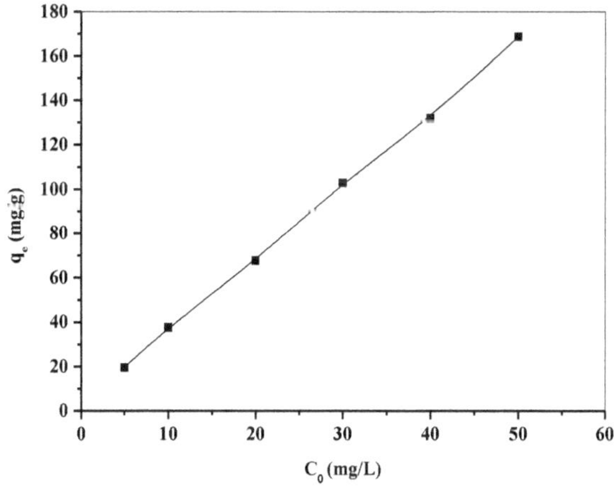

Figure VI.8. Effect of adsorbent mass on the removal efficiency and adsorption capacity of MB by MCM-bentonite: (volume of the solution = 20 mL, MCM-bentonite dosage = 5 mg, contact time = 45 min, initial pH = 7.5, room temperature = 17 °C).

To describe the interaction between adsorbate molecules and the adsorbent surface, adsorption isotherms were investigated. The analysis of experimental data using various isotherm models is of great importance, as it allows optimization of the adsorption system. Among the numerous models available, this study focused on the two most commonly applied: the Langmuir and Freundlich models.

The Langmuir model is based on the assumption of monolayer adsorption of the adsorbate on a homogeneous adsorbent surface. It does not account for surface heterogeneity and assumes that adsorption occurs only at specific, equivalent sites. The Langmuir equation is described by the following expression [11]:

$$\frac{C_e}{q_e} = \frac{1}{K_L q_{max}} + \frac{C_e}{q_{max}} \qquad (VI.3)$$

Where K_L refers to the Langmuir constant (L/mg), which is related to the adsorption energy; C_e denotes the equilibrium concentration of MB in solution (mg/L); and q_e and q_{max} correspond to the adsorption capacity at equilibrium (mg/g) and the maximum adsorption capacity of MB onto MCM-bentonite (mg/g), respectively.

In contrast, the Freundlich model considers the heterogeneity of the adsorbent surface. It describes multilayer adsorption, associated with a non-uniform energy distribution of active sites and possible interactions between adsorbed molecules. The Freundlich equation is described by [12]:

$$\log q_e = \frac{1}{n} \log C_e + \log K_F \qquad (VI.4)$$

Where K_F denotes the Freundlich adsorption constant ($(mg/g)/(mg/L)^{1/n}$), which reflects the adsorption capacity, while 1/n indicates both the adsorption intensity and the degree of surface heterogeneity of the adsorbent.

The experimental adsorption isotherms of MB onto MCM-bentonite, fitted to the Langmuir and Freundlich models, are presented in Figures VI.9(a) and VI.9(b), respectively. The calculated parameters and corresponding correlation coefficients are summarized in Table VI.2.

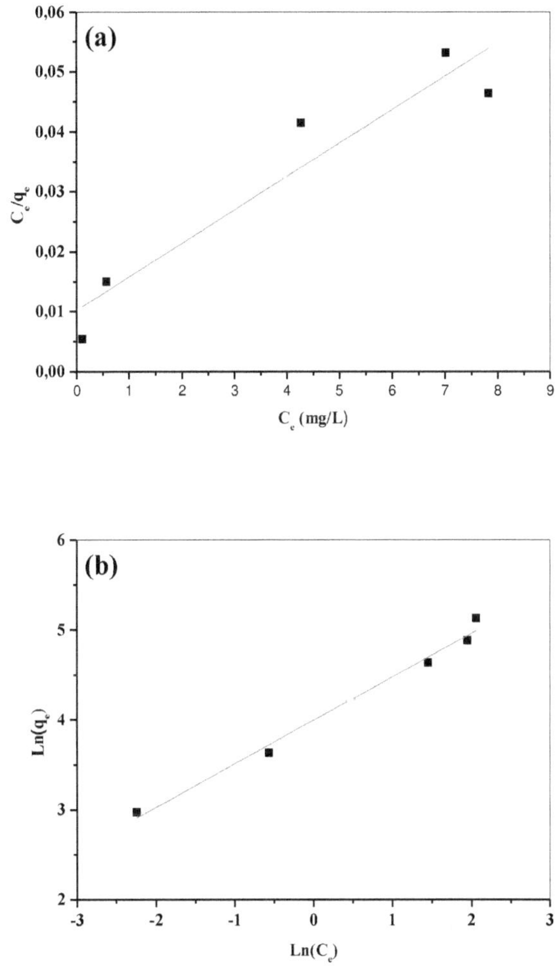

Figure VI.9. Langmuir (a) and Freundlich (b) adsorption isotherm models for MB onto MCM-bentonite.

The data reveal that the correlation coefficient for the Freundlich model (R^2 = 0.984) is considerably higher than that of the Langmuir model (R^2 = 0.879), indicating that the Freundlich model provides a better description of MB adsorption onto MCM-bentonite.

The maximum adsorption capacity (q_{max}) calculated from the Langmuir model is 179.211 mg/g for MB ions onto MCM-bentonite, highlighting the efficiency of the synthesis method in generating a large number of active adsorption sites.

The comparison of adsorption capacities of MB by various adsorbents is given in Table VI.3. The parameter 1/n, which reflects the adsorption intensity, falls within the range $0.1 < 1/n = 0.483 < 1$, indicating that the adsorption process is favorable.

Table VI.2. Langmuir and Freundlich isotherm parameters for MB adsorption onto MCM-bentonite.

Modèle	Paramètres	Valeur	R^2	Interprétation
Langmuir	qmax (mg/g)	179,211		Adsorption monocouche
	K_L (L/mg)	0,545	0,879	Sites homogènes
Freundlich	K_F	54,244		Adsorption hétérogène
	1/n	0,483	0,984	Adsorption favorable ($0 < 1/n < 1$)

A comparison of the adsorption capacities of MB using different adsorbents is presented in Table IV.3. This comparison places MCM-bentonite (179.21 mg/g) favorably among comparable adsorbents: its performance exceeds that of MCM-41 and Ti-MCM-41 prepared from TEOS as a silica source. Given its significantly lower production cost and the use of local Algerian bentonite, MCM-bentonite from Maghnia represents a particularly attractive performance–cost–local sourcing compromise.

Table VI.3. Comparison of the adsorption parameters of the MCM-bentonite adsorbent for MB with those of other MCM-41 adsorbents.

Adsorbant	Source de silice	pH	q_{max} (mg/g)	References
MCM-bentonite	Bentonite	7.5	179.21	[1]
MCM-41	TEOS	7	23	[13]
NiT50-400/MCM-41			130	

Al-MCM-41	Sodium trisilicate	8	285	[14]
MCM-41	TEOS	9	149.5	[15]
Fe-MCM-41	TEOS	9	194	
Al-MCM-41	TEOS	7	66.5	[16]
MCM-41	TEOS	7	10.95	[17]
Ti-MCM-41	TEOS	7	73.5	[18]
Zn-MCM-41	TEOS	10	166.66	[19]

VI.3.3. Effect of pH on adsorption

The pH is a crucial parameter in adsorption, as it influences both the chemical composition of the adsorbate and the surface properties of the adsorbent materials. It further governs the overall adsorption mechanism. The effect of varying initial pH values (3-10) on the adsorption capacity of MB onto MCM-bentonite (Figure 10) was examined for a contact time of 45 minutes.

The adsorption capacity of MCM-bentonite for MB increases with pH, reaching a maximum at pH 7.5. This optimal pH is particularly favorable for natural water treatment, as it generally lies within the range of 7.0-8.5. Consequently, MCM-bentonite offers an advantage over many other adsorbents that necessitate pH adjustment beyond these ambient conditions.

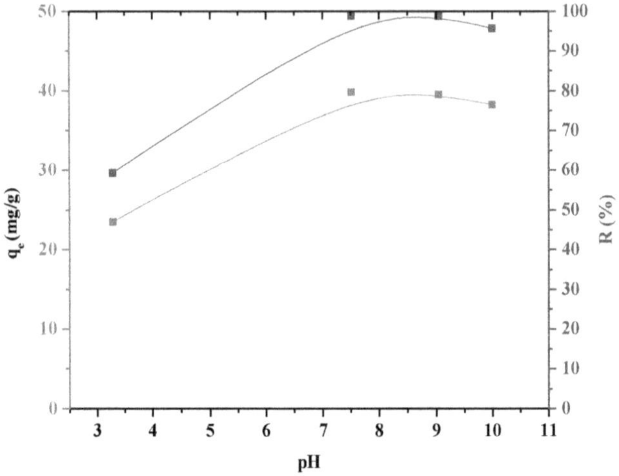

Figure VI.10. Effect of pH on the removal efficiency and adsorption capacity of MB by MCM-bentonite: (MB concentration = 10 ppm, volume of the solution = 20 mL, MCM-bentonite dosage = 5 mg, contact time = 45 min, room temperature = 17 °C).

The relatively low adsorption of MB under acidic conditions has also been reported for various adsorbents and is attributed to the competition between MB^+ ions and H^+ [20, 21]. The observed increase in MB adsorption with rising pH is explained by the deprotonation of surface silanol groups (≡Si-OH) on the adsorbent [21]. A slight decline in MB adsorption on MCM-bentonite at pH values above 7.51 was noted, likely due to silica hydrolysis occurring under alkaline conditions [22].

Several plausible adsorption mechanisms, including electrostatic interactions, hydrogen bonding, ion exchange, coordination, and acid-base interactions, have been proposed to elucidate the interactions between the solid adsorbent and dye molecules in aqueous solution. In the present study, the predominant mechanism governing MB adsorption on MCM-bentonite is electrostatic attraction (Figure VI.11). The surface silanol groups (Si-OH) of the adsorbent ionize to generate H^+ and $\equiv Si\text{-}O^-$ ions, which interact with MB^+ via electrostatic forces between oppositely charged species.

Under acidic conditions, the solution contains a high concentration of H^+ ions, which can compete with MB^+ for adsorption onto the surface of MCM-bentonite, thereby reducing the adsorption efficiency (at pH = 3.28; R = 59.24%).

However, with increasing solution pH, the ionization of surface ≡Si-OH groups is enhanced, resulting in a higher density of ≡Si-O^- groups on MCM-bentonite, which significantly improves the adsorption of MB^+ (at pH = 7.51; R = 98.94%). Furthermore, hydrogen bonding may occur between the H atoms of non-ionized surface ≡Si-OH groups on MCM-bentonite and the N atoms of MB molecules.

Consequently, both electrostatic interactions and hydrogen bonding are likely to contribute to the adsorption of MB on MCM-bentonite. These findings are in agreement with the literature [23- 27].

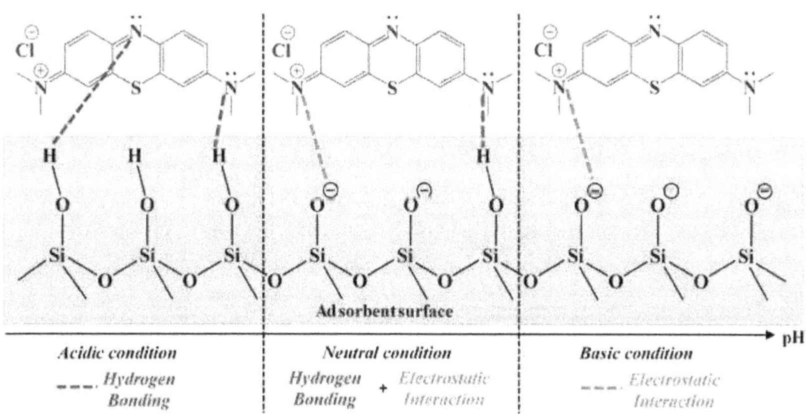

Figure VI.11. Proposed mechanism for the adsorption of MB onto MCM-bentonite.

VI.3.4. Thermodynamic parameters of adsorption

Studying the adsorption behavior as a function of temperature is crucial for elucidating the thermodynamics, spontaneity, and energetic aspects of the adsorption process, thereby providing valuable insights for optimizing system performance under varying conditions [28]. Thermodynamic parameters were determined for MB adsorption on MCM-bentonite (10 ppm MB, pH = 7.51, contact time of 45 min, 5 mg of adsorbent) at four temperatures (17 °C, 25 °C, 30 °C, and 45 °C).

The Gibbs free energy change (ΔG°) for the adsorption process, along with the enthalpy change (ΔH°) and entropy change (ΔS°), were calculated using the following equations [12]:

$$K_d = \frac{q_e}{C_e} \qquad (VI.5)$$

$$lnK_d = -\frac{\Delta H^0}{RT} + \frac{\Delta S^0}{R} \qquad (VI.6)$$

$$\Delta G^0 = \Delta H^0 - \Delta S^0 \qquad (VI.7)$$

In the aforementioned equations, R represents the ideal gas constant (8.314 $J \cdot mol^{-1} \cdot K^{-1}$), T (K) denotes the absolute temperature, and K_d is the adsorption equilibrium constant as a function of temperature.

Figure VI.12 shows that the adsorption percentage (R%) remains largely unchanged across the investigated temperature range. This suggests that the adsorption of methylene blue onto MCM-bentonite is not substantially affected by temperature variations within this interval. Accordingly, under these conditions, the interactions between MB molecules and the adsorption sites on MCM-bentonite appear to be only minimally influenced by thermal effects.

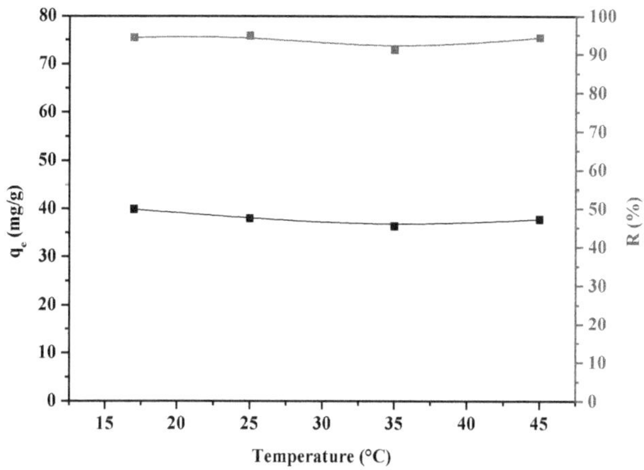

Figure VI.12. Effect of temperature on the removal efficiency and adsorption capacity of MB by MCM-bentonite: (MB concentration = 10 ppm, volume of the solution = 20 mL, MCM-bentonite dosage = 5 mg, ambient initial pH = 7.5, contact time = 45 min).

The plot of ln K_d versus (1/T) enabled the determination of ΔH° from the slope and ΔS° from the y-intercept. The thermodynamic parameters obtained (Table VI.4) corroborate the characteristics of the investigated adsorption process. The negative enthalpy value (ΔH° < 0) indicates that the adsorption is exothermic, occurring with the release of heat. Nevertheless, the slight enthalpy variation (ΔH° = -1.2252 kJ/mol) suggests interactions typical of physisorption. This finding further supports the proposed interaction mechanisms, namely hydrogen bonding between the ≡Si-OH groups of MCM-bentonite and the amine groups of MB, as well as electrostatic attractions between the negatively charged $≡SiO^-$ groups on the adsorbent surface and the positively charged MB^+ groups.

Moreover, the negative Gibbs free energy values (ΔG° < 0) indicate that the adsorption process is spontaneous at the experimental temperature. This spontaneity is further supported by the positive standard entropy change (ΔS° > 0), indicating an increase in disorder during the adsorption process. The observed rise in entropy can be attributed to the release of water molecules initially adsorbed on the adsorbent surface or surrounding the adsorbate in solution [29].

Table 4. Thermodynamic parameters for MB adsorption onto the MCM-bentonite.

Temperature (°C)	ΔG° (kJ/mol)	ΔH° (kJ/mol)	ΔS° (kJ.K/mol)
17	-9,8716		
25	-10,1100	-1,2252	0,0298
35	-10,2590		
45	-10,7060		

VI.4. Conclusion du Chapitre VI

The development of cost-effective and sustainable adsorbents remains a critical challenge in wastewater treatment. In this context, a mesoporous silica MCM-41 adsorbent was successfully synthesized from natural Algerian bentonite using an alkaline fusion approach as an alternative silica source. X-ray diffraction analysis confirmed the complete removal of the original bentonite crystalline phases and the formation of a well-ordered hexagonal mesostructure characteristic of MCM-41.

The synthesized material exhibited excellent adsorption performance toward methylene blue (MB), achieving a removal efficiency of 98.94% at the natural solution pH (7.51). Rapid adsorption within the first 5 min highlights the strong affinity between the adsorbent surface and MB molecules. Kinetic data were best described by the pseudo-second-order model, while equilibrium results followed the Freundlich isotherm, indicating heterogeneous adsorption behavior. The maximum adsorption capacity, estimated from the Langmuir model, reached 179.24 $mg \cdot g^{-1}$, surpassing values reported for several MCM-41 materials derived from conventional silica sources.

The adsorption mechanism is primarily governed by electrostatic interactions between negatively charged silanol groups ($\equiv SiO^-$) and MB^+ species, with additional contributions from hydrogen bonding. Thermodynamic analysis revealed a spontaneous and exothermic process ($\Delta G° < 0$, $\Delta H° < 0$), with low enthalpy values consistent with physisorption and a positive $\Delta S°$ reflecting increased interfacial disorder.

Overall, these findings demonstrate that bentonite-derived MCM-41 is an efficient, low-cost, and sustainable adsorbent, offering a promising alternative for the removal of cationic dyes and advancing the development of environmentally friendly wastewater treatment technologies.

VI.5. Prospects for the Treatment of Algerian Industrial Wastewater

The results obtained on the adsorption of methylene blue open up concrete prospects for the treatment of Algerian industrial effluents, particularly those from textile industries in the Tlemcen region and mining effluents related to bentonite exploitation itself. Several development directions are envisaged:

Heavy metal adsorption: Functionalization of MCM-bentonite with amino groups (APTES grafting) or thiol groups (MPTMS grafting) could selectively enhance the affinity for Pb^{2+}, Cu^{2+}, Cd^{2+}, and Hg^{2+} ions, which are major pollutants in industrial effluents.

Anionic dye adsorption: For anionic dyes (methyl orange, Congo red), cationic modification of the MCM-bentonite surface (through grafting of protonated amino groups) would be required to reverse the surface charge and enable adsorption.

Regeneration and reuse: Regeneration studies using washing with dilute NaOH (0.1 M) or mild thermal treatment (150°C, 2 h) would make it possible to evaluate the number of adsorption–regeneration cycles achievable with MCM-bentonite, a key factor in assessing the economic attractiveness of the process at pilot scale.

VI.6. References of Chapter VI

1. Mesli W.M., Merad H.Z. (2025). Traitement des eaux par adsorption sur MCM-41 synthétisé à partir de la bentonite de Maghnia. Mémoire de Master, Spécialité Chimie Macromoléculaire, Département de Chimie, Université Abou Bekr Belkaïd de Tlemcen.

2. Liu Y., Li C., Peyravi A., Sun Z., Zhang G., Rahmani K., Zheng S., Hashisho Z. (2021). Mesoporous MCM-41 derived from natural Opoka and its application for organic vapors removal. *Journal of Hazardous Materials*, 408, 124911.

3. Hassan H.M., El-Sayed M.A., El-Maghraby M.A., El-Sayed A.F., El-Shafey A.M.M. (2025). Nanosized MCM-41 silica from rice husk and its application for the removal of organic dyes from water. *RSC Advances*, 15, 2545–2553.

4. Costa J.A.S., de Jesus R.A., Santos D.O., Mano J.F., Cruz L.P. (2020). Recent progresses in the adsorption of organic, inorganic, and gas compounds by MCM-41-based mesoporous materials. *Microporous and Mesoporous Materials*, 291, 109698.

5. Moukahhal K. (2020). Synthèse et mise en forme d'objets poreux hiérarchisés composés de nanozéolithes. Ph.D. thesis, Université de Haute-Alsace, Mulhouse, France.

6. Rouquerol J., Rouquerol F., Sing K. (2012). *Adsorption by Powders and Porous Solids: Principles, Methodology and Applications*. Academic Press.

7. Lagergren S. (1898). About the theory of so-called adsorption of solution substances. *K. Sven. Vetenskapsakad. Handl.*, 24, 1–39.

8. Ho Y.S., McKay G. (1999). Pseudo-second order model for sorption processes. *Process Biochemistry*, 34, 451–465.

9. Servant L., Le Bourdon G., Buffeteau T. (2011). Comprendre la spectroscopie infrarouge : principes et mise en œuvre. *Photoniques*, 53, 68–71.

10. Luo B., Zhang M., Zhou N. et al. (2025). Mechanism study of a novel modified mesoporous silicon-based NiT50/MCM-41 adsorbent for the selective adsorption of methylene blue from water. *RSC Advances*, 15, 5737–5749.

11. Dhal J.P., Dash T., Hota G. (2020). Iron oxide impregnated mesoporous MCM-41: synthesis, characterization and adsorption studies. *Journal of Porous Materials*, 27, 205–216.

12. Drweesh S.A., Fathy N.A., Wahba M.A. et al. (2016). Equilibrium, kinetic and thermodynamic studies of Pb(II) adsorption from aqueous solutions on HCl-treated Egyptian kaolin. *Journal of Environmental Chemical Engineering*, 4, 1674–1684.

13. Luo B., Zhang M., Zhou N. et al. (2025). Mechanism study of a novel modified mesoporous silicon-based NiT50/MCM-41 adsorbent for the selective adsorption of methylene blue from water. *RSC Advances*, 15, 5737–5749.

14. Rashwan W.E., Abou-El-Sherbini K.S., Wahba M.A. et al. (2020). High stable Al-MCM-41: structural characterization and evaluation for removal of methylene blue from aqueous solution. *Silicon*, 12, 2017–2029.

15. Dhal J.P., Dash T., Hota G. (2020). Iron oxide impregnated mesoporous MCM-41: synthesis, characterization and adsorption studies. *Journal of Porous Materials*, 27, 205–216.

16. Eftekhari S., Habibi-Yangjeh A., Sohrabnezhad S. (2010). Application of AlMCM-41 for competitive adsorption of methylene blue and rhodamine B: thermodynamic and kinetic studies. *Journal of Hazardous Materials*, 178, 349–355.

17. Xiao X., Zhang F., Feng Z., Deng S., Wang Y. (2015). Adsorptive removal and kinetics of methylene blue from aqueous solution using NiO/MCM-41 composite. *Physica E: Low-Dimensional Systems and Nanostructures*, 65, 4–12.

18. Zanjanchi M., Golmojdeh H., Arvand M. (2009). Enhanced adsorptive and photocatalytic achievements in removal of methylene blue by incorporating tungstophosphoric acid-TiO2 into MCM-41. *Journal of Hazardous Materials*, 169, 233–239.

19. Hachemaoui M., Boukoussa B., Mokhtar A. et al. (2020). Dyes adsorption, antifungal and antibacterial properties of metal loaded mesoporous silica: Effect of metal and calcination treatment. *Materials Chemistry and Physics*, 256, 123704.

20. Dang W.X., Han S.H., Xu J. et al. (2004). Effect of formamide on the morphologies of ordered mesoporous silica. *Chinese Journal of Inorganic Chemistry*, 20(6), 679–682.

21. Drweesh S.A., Fathy N.A., Wahba M.A. et al. (2016). Equilibrium, kinetic and thermodynamic studies of Pb(II) adsorption from aqueous solutions on HCl-treated Egyptian kaolin. *Journal of Environmental Chemical Engineering*, 4(2), 1674–1684.

22. Abou-El-Sherbini K.S., Schiel D., Stosch R. et al. (2011). Stabilization of quercetin-functionalized silica gel against hydrolysis by blocking silanol groups with TiO_2 or ZrO_2 and its application for the removal of Hg(II). *Journal of Sol-Gel Science and Technology*, 57, 57–67.

23. Zhou C., Gao Q., Luo W. et al. (2015). Preparation, characterization and adsorption evaluation of spherical mesoporous Al-MCM-41 from coal fly ash. *Journal of the Taiwan Institute of Chemical Engineers*, 52, 147–157.

24. Wang W., Tian G., Wang D. et al. (2016). All-into-one strategy to synthesize mesoporous hybrid silicate microspheres from naturally rich red palygorskite clay as high-efficient adsorbents. *Scientific Reports*, 6, 39599.

25. Liu M., Hou L.A., Xi B. et al. (2013). Synthesis, characterization, and mercury adsorption properties of hybrid mesoporous aluminosilicate sieve prepared with fly ash. *Applied Surface Science*, 273, 706–716.

26. Han X., Wang Y., Zhang N. et al. (2021). Facile synthesis of mesoporous silica derived from iron ore tailings for efficient adsorption of methylene blue. *Colloids and Surfaces A: Physicochemical and Engineering Aspects*, 617, 126391.

27. Ma Z., Liao H., Pan Z. et al. (2022). Insights into coproduction of silica gel via desulfurization of steel slag and silica gel adsorption performance. *ACS Omega*, 7(24), 21062–21074.

28. Demirtaş H., Taşar Ş., Kaya F. et al. (2022). Sorption of BB41 dye molecules using chitosan based particles from aqueous solutions: A kinetic and thermodynamic evaluation. *Journal of Environmental Chemical Engineering*, 10(4), 108062.

29. Azhagiya Singam E.R., Zhang Y., Magnin G. et al. (2019). Thermodynamics of adsorption on graphenic surfaces from aqueous solution. *Journal of Chemical Theory and Computation*, 15(2), 1302–1316.

GENERAL CONCLUSION

All of this work converges toward a clear scientific and societal objective: to valorize Algerian bentonite from Maghnia, a naturally abundant and low-cost mineral resource, for the synthesis of functional mesoporous Al-MCM-41 materials, and to exploit these materials in high-impact applications, namely green catalysis for the synthesis of bioactive molecules and wastewater treatment.

The main original contributions of this research program can be summarized into five distinct achievements:

Demonstration of the feasibility of MCM-41 synthesis from Algerian bentonite:

For the first time, it was shown that structurally well-ordered aluminosilicate MCM-41 can be synthesized from Maghnia bentonite without the addition of pure chemical silica or alumina sources. This fundamental result formed the basis for all subsequent studies.

Rigorous optimization of synthesis parameters:

The relationships between synthesis parameters and material properties were established for MCM-bentonite. Optimal conditions were identified (pH = 9.5, T = 100°C, 48 h), and a general pH–temperature correlation rule was formulated. This systematic study provides a conceptual framework for reproducing and improving the synthesis process.

Quantitative characterization of Lewis acidity:

A first comparative quantitative study of Lewis acidity between standard Al-MCM-41 and Al-MCM-bentonite was provided, linking acidity differences (factor ~8) to the amount of aluminum incorporated into the hexagonal phase (Si/Al = 13.47 vs 43.64). This analysis offers a mechanistic understanding of the catalytic limitations of MCM-bentonite.

Original green catalytic application:

The efficiency of Al-MCM-41 was demonstrated as a heterogeneous catalyst for the solvent-free synthesis of formamidine derivatives at room temperature, with higher yields (93–99%) and significantly reduced reaction times (5–12-fold decrease) compared to conventional thermal methods. Recyclability over six cycles confirms the catalyst's sustainability.

Fifth contribution — Environmental application for water treatment:

MCM-bentonite proved to be an efficient adsorbent (q_{max} = 179.2 mg/g for methylene blue), showing competitive performance compared to MCM-41 prepared from TEOS as a silica source.

Overall, the work presented in this thesis forms a coherent body around a central question: how can Algerian bentonite, an abundant and underutilized natural mineral resource, be transformed into high-value functional Al-MCM-41 mesoporous materials, and how can these materials be exploited in catalytic and environmental applications aligned with sustainable development principles? Addressing this question required a rigorous stepwise scientific approach, from synthesis and optimization to applications, supported by in depth physicochemical characterization.

The overall process pathway can be represented as follows: (Maghnia Bentonite) → [Alkaline activation: NaOH fusion] → (Aluminosilicate supernatant) → [Hydrothermal synthesis: CTABr, 100°C, pH = 9.5, 48 h] → (Calcined Al-MCM-41/bentonite) → [Applications: heterogeneous catalysis + adsorption]. This scheme summarizes the added value generated at each stage and highlights the coherence of the approach.

A critical analysis of the obtained results allows clear identification of both strengths and limitations of the bentonite-based Al-MCM-41 synthesis strategy.

Main strengths: The first strength lies in the unequivocal demonstration of the feasibility of synthesizing well-ordered aluminosilicate MCM-41 from Maghnia bentonite without external silica or alumina sources, validating the concept and enabling low-cost production. The second strength is the remarkably well-developed hexagonal ordering of Al-MCM-bentonite, superior to standard Al-MCM-41 at comparable aluminum contents, attributed to naturally occurring cations acting as bridging agents. The third strength is the demonstrated catalytic activity of Al-MCM-bentonite in green formamidine synthesis under optimized conditions, despite its lower Lewis acidity. The fourth strength is the significantly reduced production cost (≈50% reduction in raw material cost) and the use of a local Algerian resource, providing strong economic and strategic advantages.

Identified limitations:

The main limitation is the lower specific surface area (494–674 m^2/g vs 983–1154 m^2/g for the standard material) and lower Lewis acidity (172.8 vs 1414.8 µmol/g) of Al-MCM-bentonite. These limitations stem directly from the lower aluminum incorporation in the hexagonal phase (Si/Al = 43.64 vs 13.47), linked to the alkaline fusion conditions. A second limitation is the persistent presence of an amorphous phase fraction in Al-MCM-bentonite, which is difficult to eliminate without fundamentally modifying the synthesis protocol. These limitations define clear improvement pathways and constitute the priority perspectives of this work.

A major strength of this research program is the strong coherence between the results obtained from different characterization techniques and application tests. This coherence is observed at several levels:

Structure–acidity–activity coherence:

The quantitative relationship between the Si/Al ratio in the hexagonal phase (EDX-SEM), Lewis acid site density (FTIR-pyridine), and catalytic activity in test reactions (allylation and formamidine synthesis) is fully consistent: Si/Al(hex) = 43.64 → Lewis acidity = 51.7 µmol/g → allylation yield = 20% (24 h) for Al-MCM-bentonite; Si/Al(hex) = 13.47 → Lewis acidity = 969.2 µmol/g → allylation yield = 99% (4 h) for standard Al-MCM-41.

Synthesis–texture–adsorption coherence:

The specific surface area of MCM-bentonite (494 m^2/g) is directly linked to the quality of hexagonal ordering (a_0 = 44.00 Å, Dp = 38 Å) and synthesis conditions (pH = 9.5, T = 100°C). This surface area governs methylene blue adsorption capacity (qmax = 179.21 mg/g), consistent with literature values for materials of similar textural properties.

In conclusion, this work demonstrates the feasibility and scientific relevance of valorizing Algerian bentonite from Maghnia for the synthesis of functional mesoporous materials. The resulting Al-MCM-41 is a multifunctional, cost-effective, and environmentally friendly heterogeneous catalyst and adsorbent, with clearly established potential in green catalysis and wastewater treatment. The perspectives outlined open up promising research avenues for future work within our group and the national scientific community.

LIST OF ABBREVIATIONS AND SYMBOLS

Symbol / Abbreviation	Full meaning	Unit
a_0	Hexagonal lattice parameter of MCM-41	Å
Al-MCM-41	Aluminosilicate MCM-41 (Al-incorporated)	—
ATD / ATG	Differential Thermal Analysis / Thermogravimetric Analysis (DTA/TGA)	—
BET	Brunauer–Emmett–Teller method (specific surface area)	m^2/g
BJH	Barrett–Joyner–Halenda method (pore size distribution)	Å
C	BET constant (gas–surface interaction energy)	—
CB	Concentration of Brønsted acid sites	µmol/g
CEC	Cation exchange capacity	meq/100 g
CL	Concentration of Lewis acid sites	µmol/g
CMC	Critical micelle concentration	mmol/L
CTABr	Cetyltrimethylammonium bromide (C_{16} surfactant)	—
d_{100}	Interplanar spacing (100) of MCM-41	Å
DMF-DMA	N,N-Dimethylformamide dimethyl acetal	—
Dp	Pore diameter (BJH)	Å
DRX	X-ray diffraction (XRD)	—
EDX / EDS	Energy-dispersive X-ray spectroscopy	—
ESSA-T	Higher School of Applied Sciences of Tlemcen	—
E-factor	Environmental factor (waste mass / product mass)	—
FTIR	Fourier Transform Infrared Spectroscopy	cm^{-1}
IUPAC	International Union of Pure and Applied Chemistry	—
LCM	Materials Chemistry Laboratory (University Oran 1)	—
LEM	Environment and Mineralogy Laboratory (Nancy, France)	—
M41S	Mobil mesoporous materials family (1992)	—

MCM-41	Mobil Crystalline Material No. 41	—
SEM (MEB)	Scanning Electron Microscopy	—
TEM (MET)	Transmission Electron Microscopy	—
NH_4OH	Ammonium hydroxide	—
NaOH	Sodium hydroxide	—
P/P_0	Relative pressure (gas adsorption)	—
PMO	Periodic Mesoporous Organosilica	—
qe	Equilibrium adsorption capacity	mg/g
qmax	Maximum adsorption capacity (Langmuir)	mg/g
R^2	Coefficient of determination	—
SBET	BET specific surface area	m^2/g
Si-MCM-41	Purely siliceous MCM-41 (without Al)	—
Si/Al	Silicon/aluminium molar ratio	—
TEOS	Tetraethyl orthosilicate $Si(OC_2H_5)_4$	—
TMAOH	Tetramethylammonium hydroxide	—
Vmeso	Mesopore volume	cm^3/g
Wt	Mesopore wall thickness (a_0 – Dp)	Å
ΔG°	Standard Gibbs free energy of adsorption	kJ/mol
ΔH°	Standard enthalpy of adsorption	kJ/mol
ΔS°	Standard entropy of adsorption	J/mol·K

APPENDICES

A.1. Complete XRD Data for All Materials

A.1.1. Structural Parameters Before and After Calcination

Table VII.1 summarizes the complete set of structural parameters extracted from the small-angle XRD diffractograms for all MCM-41 materials synthesized and characterized within the framework of the work presented in this book. These data constitute the basis for all structural comparisons discussed in the previous chapters.

Tableau VII.1. Complete XRD parameters before calcination (NC) and after calcination (C) for all investigated MCM-41 materials. std = standard (pure reagents); Ben = bentonite.

Materials	Synthesis pH	Synthesis T (°C)	2θ(100) (NC) (°)	d_{100} (NC) (Å)	a_0 (NC) (Å)	2θ(100) (C) (°)	d_{100} (C) (Å)	a_0 (C) (Å)	Δa_0 (%)
Si-MCM-std	—	100	2.21	39.94	46.12	2.45	36.03	41.60	9.8
Al-MCM-std	—	100	2.04	43.27	49.96	2.32	38.05	43.94	12.0
MCM-Ben (Maghnia)	9.5	100	2.27	38.89	44.91	2.29	38.55	44.51	0.9
MCM-Ben-pH8.5	8.5	80	2.42	42.36	48.92	2.80	36.61	42.28	—
MCM-Ben-pH9.5	9.5	100	2.32	44.18	51.08	2.69	38.10	44.00	—
MCM-Ben-pH11.5	11.5	130	2.56	40.04	46.24	2.63	38.98	45.02	—
MCM-Volclay	—	100	2.20	40.10	46.30	2.23	39.60	45.72	1.3

A.1.2. Contraction Rate and Thermal Stability

The unit-cell contraction rate during calcination::

$$\Delta a_0 = [a_0(NC) - a_0(C)] / a_0(NC) \times 100\ \%)$$

Is an important indicator of the thermal stability of the materials. Al-MCM-bentonite exhibits the lowest value ($\Delta a_0 = 0.9\%$), indicating excellent wall cohesion during calcination. This behavior is attributed to the presence of divalent cations (Ca^{2+}, Mg^{2+}) naturally present in bentonite, which are incorporated into the mesoporous material and reinforce the bonding between silicate chains within the pore walls. In comparison, Al-MCM-standard exhibits $\Delta a_0 = 12.0\%$, a significantly higher value reflecting a looser and less cross-linked siliceous network.

A.2. Detailed BET/BJH and Textural Data

Tableau VII.2. Complete textural data (BET and BJH) for all investigated MCM-41 materials. Porosity calculated using Vmeso/(Vmeso + 1/ρ amorphous silica) × 100%.

Materials	SBET (m^2/g)	Vm (cm^3/g)	C BET	Vmeso (cm^3/g)	Dp BJH (Å)	Wt (Å)	Porosity (%)
Si-MCM-std	1154	265,2	112	0,84	34,0	7,6	62
Al-MCM-std	983	226,1	105	0,76	35,6	8,4	59
Al-MCM-bentonite	674	154,9	87	0,42	35,6	8,9	51
MCM-Ben-pH=8,5	373	85,8	72	0,59	33,4	8,9	55
MCM-Ben-pH=9,5	494	113,5	98	0,72	38,0	6,0	59
MCM-Ben-pH=11,5	417	95,9	88	0,60	35,7	9,4	57

A.3. Complete Quantitative Acidity Data

Table VII.3 presents the complete concentrations of Brønsted acid sites (CB) and Lewis acid sites (CL) for the two Al-MCM-41 materials investigated in Chapter IV, as determined by pyridine-FTIR at four successive desorption temperatures (50, 150, 250, and 350°C). These data make it possible to evaluate not only the quantity but also the strength distribution of the acid sites.

Tableau VII.3. Concentrations of Lewis and Brønsted acid sites (µmol/g) determined by pyridine-FTIR at different desorption temperatures for Al-MCM-standard and Al-MCM-bentonite.

Site Type	Material	50°C	150°C	250°C	350°C	Σ Total	CL/CB
Lewis CL (µmol/g)	Al-MCM-standard	203.49	247.42	259.74	258.53	969.18	2.17
	Al-MCM-bentonite	4.34	12.07	16.17	19.07	51.65	0.42
Brønsted CB (µmol/g)	Al-MCM-standard	92.13	124.79	142.78	85.89	445.59	—
	Al-MCM-bentonite	26.79	30.10	31.20	33.03	121.12	—
Total CL+CB (µmol/g)	Al-MCM-standard	295.62	372.21	402.52	344.42	1414.77	—
	Al-MCM-bentonite	31.13	42.17	47.37	52.10	172.77	—

The evolution of acid-site concentrations with desorption temperature also provides valuable mechanistic information. For Al-MCM-standard, CL increases from 203.49 µmol/g at 50°C to 259.74 µmol/g at 250°C before slightly decreasing at 350°C. This initial increase is attributed to the desorption of weakly physisorbed pyridine molecules that masked Lewis sites, progressively revealing the active sites. For Al-MCM-bentonite, the similar but lower-amplitude evolution (4.34 → 19.07 µmol/g) confirms the presence of a limited number of Lewis sites with moderate strength.

A.4. Complete Catalytic Data

A.4.1. Table of All Yields — Formamidine Synthesis

Table VII.4 compiles the complete yield data for the eight synthesized formamidine derivatives, with systematic comparison between the catalytic method (Al-MCM-41, optimized conditions) and the conventional thermal method (without catalyst, reflux in p-xylene or another solvent).

Tableau VII.4. Complete yield and reaction-time data for the synthesis of formamidines 2a–2h. THBT = 4,5,6,7-tetrahydrobenzo[b]thiophene. Cat. = Al-MCM-41 10 wt%, room temperature, solvent-free. Therm. = without catalyst, reflux in p-xylene.

Formamidine	Amine	Catalytic yield (%)	Catalytic time (min)	Thermal yield (%)	Thermal time (min)	Δt (factor)	Δyield (%)
2a	3-Amino-4-Ph-6-p-tolylpyr.	95	30	75	210	7.0×	+20
2b	3-Amino-4,6-diphenylpyr.	97	30	87	270	9.0×	+10
2c	3-Amino-4-(4-Cl)-6-p-tolyl	97	30	90	270	9.0×	+7
2d	3-Amino-6-(4-Br)-4-(4-Cl)	98	45	79	225	5.0×	+19
2e	3-Amino-4-(4-F)-6-(3-OMe)	93	30	88	210	7.0×	+5
2f	3-Amino-6-(4-F)-4-(thioph.)	96	30	82	270	9.0×	+14
2g	Methyl 2-amino-THBT	99	60	78	720	12.0×	+21
2h	3-Amino-THBT	96	60	78	300	5.0×	+18

B.1. Complete Protocol for the Synthesis of Al-MCM-41 from Bentonite

B.1.1. Required Equipment

- Alumina crucible (100 mL), tubular furnace, analytical balance (0.1 mg precision).
- Glass beakers (100 mL, 250 mL), magnetic hotplate stirrer, calibrated pH meter.
- Stainless-steel autoclave with Teflon liner (100 mL), ventilated drying oven.
- Centrifuge (4000 rpm), Büchner frit funnel, vacuum pump.
- Tubular furnace for calcination, programmable temperature controller.

B.1.2. Chemicals

- Natural bentonite from Maghnia (ground to < 100 μm).
- NaOH pellets (≥ 97%).
- CTABr (Merck, ≥ 99%), NH_4OH (25%).
- Glacial acetic acid diluted to 1 M.
- Demineralized water (conductivity < 5 μS/cm).

B.1.3. Detailed Experimental Procedure

Step 1 — Preparation of Fused Bentonite

Weigh 10.0 g of bentonite and 12.0 g of NaOH. Thoroughly mix the powders in an alumina crucible using a spatula. Place the crucible in a tubular furnace and program the thermal cycle as follows: heating at 5°C/min up to 550°C, holding for 60 minutes, followed by free cooling. Allow the sample to cool to room temperature inside the furnace. Recover the fused solid (whitish-gray in color) and finely grind it using a porcelain mortar.

Step 2 — Extraction of the Aluminosilicate Supernatant

Transfer the ground fused solid into a 250 mL beaker. Add 40 mL of demineralized water (fused solid/water mass ratio = 1/4). Stir on a magnetic stirrer at 400 rpm for 24 h at room temperature. Centrifuge the suspension at 4000 rpm for 20 min. Carefully collect the clear supernatant using a Pasteur pipette. Measure the pH (typically pH ≈ 12–13) and record the recovered volume (typically 35–40 mL).

Step 3 — Preparation of the Synthesis Gel

In a 100 mL beaker, weigh 0.867 g of CTABr and dissolve it in 15 mL of demineralized water. Add 0.75 g of NH_4OH under stirring. Slowly pour the 40 mL of supernatant into this solution under vigorous stirring. Continue stirring for 30 min. Measure the pH: if pH > 10, add 1 M acetic acid dropwise until pH = 9.5 ± 0.1 (approximately 2–5 mL depending on the basicity of the supernatant). Acid addition must be slow and gradual to avoid local silica precipitation.

Step 4 — Hydrothermal Treatment

Quantitatively transfer the gel into the Teflon liner of the autoclave. Close the autoclave by tightening the screws uniformly. Place the autoclave in a drying oven preheated to 100°C. Maintain at 100°C for 48 h without stirring.

Step 5 — Recovery of the Product

Open the autoclave after cooling to room temperature (**CAUTION:** do not open while hot). Filter the product using a Büchner frit funnel under vacuum. Wash thoroughly with demineralized water (5 × 50 mL) until the filtrate reaches neutral pH. Dry the filter cake at 100°C for 24 h in a ventilated oven. Weigh the dried non-calcined product (typically 1.8–2.2 g, corresponding to a yield of 65–80% of non-calcined MCM-41).

Step 6 — Calcination

Transfer the dried product into a silica crucible. Program the calcination cycle as follows: heating at 2°C/min up to 200°C (holding for 30 min to remove residual moisture), then heating at 2°C/min up to 550°C (holding for 12 h under an air flow of 50 mL/min), followed by slow cooling at 1°C/min down to 200°C and then free cooling. The calcined material (white to off-white) is weighed (typically 0.9–1.1 g) and stored in a sealed vial protected from moisture.

B.2. FTIR-Pyridine Characterization Protocol

Pellet Preparation

Weigh 2 ± 0.05 mg of finely ground calcined sample. Thoroughly mix with 200 mg of anhydrous KBr (previously dried for 12 h at 200°C). Compress under 10 tons for 2 min using a hydraulic press. Obtain a translucent pellet of 13 mm diameter. Precisely weigh the pellet (m ± 0.1 mg) for the quantitative calculation of acid-site concentrations.

Degassing

Place the pellet in the variable-temperature FTIR cell. Connect the cell to the vacuum line (rotary pump + turbomolecular pump). Degas progressively under the following conditions: 100°C (1 h), 200°C (1 h), 350°C (1 h), and 450°C (12 h, final vacuum $< 10^{-3}$ Torr). Record the FTIR spectrum of the degassed material (background spectrum, baseline).

Pyridine Adsorption

Introduce distilled pyridine vapor (saturated vapor pressure at 25°C = 20 Torr) into the cell at 150°C for 15 min. Desorb excess pyridine at 150°C under vacuum for 15 min. Record the FTIR spectrum. Repeat desorption successively at 250°C (15 min) and 350°C (15 min), recording the spectrum after each step.

Concentration Calculations

Integrate the bands at 1456 cm^{-1} (Lewis sites) and 1545 cm^{-1} (Brønsted sites) at each temperature.

Apply the Emeis equations:

- CB (µmol/g) = 1,88 × A_{1545} × S / (εB × m)
- CL (µmol/g) = 1,42 × A_{1456} × S / (εL × m)
- Avec εB = 0,73 cm·μmol^{-1} et εL = 1,11 cm·μmol^{-1}.

Term	Definition
Brønsted acidity	Site capable of donating an H^+ proton. FTIR-pyridine band at 1545 cm^{-1}.
Lewis acidity	Site capable of accepting an electron pair. FTIR-pyridine band at 1456 cm^{-1}.
Physical adsorption	Reversible fixation of molecules on a surface through van der Waals forces.
Alkaline activation	Clay + NaOH fusion at 550°C to dissolve and extract Si and Al species.
Al-MCM-41	Aluminosilicate MCM-41 containing Al through isomorphic substitution of Si.
Calcination	Thermal treatment (550°C, 12 h, air) used to remove the organic surfactant from MCM-41.
CTABr	Cetyltrimethylammonium bromide, $C_{19}H_{42}NBr$: structuring surfactant for MCM-41.
Green chemistry	Design of chemical processes minimizing environmental and health impacts.
DMF-DMA	N,N-Dimethylformamide dimethyl acetal: reagent used for formamidine synthesis.
Atom economy	Ratio MW(product)/Σ MW(reactants) × 100%: indicator of green chemistry efficiency.
E-factor	Waste mass / product mass: indicator of environmental impact according to Sheldon.
Formamidine	Compound R-N=CH-NR'$_2$: precursor of nitrogen-containing heterocycles and NH_2 protecting group.
FTIR	Fourier Transform Infrared Spectroscopy. Resolution: 4 cm^{-1}.
Alkaline fusion	Bentonite + NaOH mixture heated at 550°C to extract soluble Si/Al species.
Langmuir isotherm	Monolayer adsorption model: $C_e/q_e = C_e/q_{max} + 1/(q_{max} \cdot K_L)$.
Type IV isotherm	Isotherm characteristic of mesoporous materials (IUPAC).

MCM-41	Mobil Crystalline Material No. 41: 2D hexagonal mesoporous material (Mobil, 1992).
Mesoporous	Pores with diameters of 2–50 nm (IUPAC classification, 1985).
TEM	Transmission Electron Microscopy: direct visualization of nanometric structure.
Montmorillonite	2:1 (TOT) clay mineral constituting the main component of bentonite.
Unit cell parameter a_0	$a_0 = 2d_{100}/\sqrt{3}$: dimension of the hexagonal unit cell of MCM-41.
Pseudo-second-order	Adsorption kinetic model: $t/qt = 1/(k_2qe^2) + t/qe$ (Ho and McKay).
Si/Al ratio	Indicator of aluminum content: low Si/Al = high acidity in MCM-41.
SBET	BET specific surface area (Brunauer–Emmett–Teller) expressed in m^2/g.
Surfactant	Amphiphilic molecule (polar head + hydrophobic tail) structuring MCM-41.
Hydrothermal treatment	Chemical reaction at T > 100°C under autogenous pressure in a sealed autoclave.
Vmeso	Mesoporous volume: total volume of the mesopores in a material (cm^3/g).
Wt (thickness)	Mesopore wall thickness: $Wt = a_0 - Dp$ (Å).

Publisher: Eliva Press Global Ltd

Email: info@elivapress.com

Eliva Press is an independent international publishing house established for the publication and dissemination of academic works all over the world. Company provides high quality and professional service for all of our authors.

Our Services:
Free of charge, open-minded, eco-friendly, innovational.

-Free standard publishing services (manuscript review, step-by-step book preparation, publication, distribution, and marketing).

-No financial risk. The author is not obliged to pay any hidden fees for publication.

-Editors. Dedicated editors will assist step by step through the projects.

-Money paid to the author for every book sold. Up to 50% royalties guaranteed.

-ISBN (International Standard Book Number). We assign a unique ISBN to every Eliva Press book.

-Digital archive storage. Books will be available online for a long time. We don't need to have a stock of our titles. No unsold copies. Eliva Press uses environment friendly print on demand technology that limits the needs of publishing business. We care about environment and share these principles with our customers.

-Cover design. Cover art is designed by a professional designer.

-Worldwide distribution. We continue expanding our distribution channels to make sure that all readers have access to our books.

-Marketing tools. We provide marketing tools such as banners, paid advertising, social media promotion and more.

www.ingramcontent.com/pod-product-compliance
Lightning Source LLC
LaVergne TN
LVHW022101200926
848481LV00003B/1013
* 9 7 8 9 9 9 9 3 4 4 9 6 8 *